I0105737

Language: English

These materials are designed to assist you in learning about Hope. They should not be used for medical advice, counseling, or other health-related services. iFred, The Shine Hope Company and Kathryn Goetzke do not endorse or provide any medical advice, diagnosis, or treatment. The information provided herein should not be used for the diagnosis or treatment of any medical condition and cannot be substituted for the advice of physicians, licensed professionals, or therapists who are familiar with your specific situation. Consult a licensed medical professional, or call 911, if you are in need of immediate assistance.

ISBN: 978-1-954531-15-4

For more information, please contact kathryngoetzke@theshinehopecompany.com.

We would like to thank the following people for their contribution to our programs:

This program would not be possible without the brilliant leadership, support, and commitment to Hope by:
Myron L. Belfer MD, MPA, Harvard Catalyst
Myron is Professor of Psychiatry in the Department of Psychiatry, Children's Hospital Boston, Harvard Medical School, and Senior Associate in Psychiatry at the Children's Hospital of Boston. Dr. Belfer is a Champion for Hope.

Kathryn Goetzke MBA, Author, Creator
Contributors: Taylor Steed, Katharine Lee-Kramer, Veronica O'Brien
Sarah Mellen, Mic Fariscal, Anna Termulo Montances and **Naneth Samoya-Jumawid**

To our advisors, Hope contributors, and experts:
Dr. Edward Barksdale, Dr. Frank Gard Jameson, Mayor Hillary Schieve, Kristy L. Stark M.A., Ed.M., BCBA, Karen Kirby PhD, MSc, BSc, C.Psychol, AfBPS, SFHEA, Ulster University, **Marie Dunne and the Northern Ireland team** that helped plant the seeds for this work.

Pioneers in early Hope Science including **Dr. Crystal Bryce, Dr. Dan Tomasulo, Dr. Chan Hellman, Dr. Matthew Gallagher, Dr. Jennifer Cheavens and the late Dr. Shane Lopez.**

The Hopeful Minds Advisory Board

iFred Board of Directors:
Tom Dean, Susan Minamyer, Jim Link, Dr. John Grohol, Kathryn Goetzke, Dr. Mindy Magrane

Some of our early funders: Sutter Health, Anthem, The Gordon Family Giving Fund of the Parasol Tahoe Community Foundation, The Shine Hope Company, and The Mood Factory.

IN SPECIAL RECOGNITION
Susan Minamyer, whose unconditional love, support, encouragement, faith, and brilliance planted and watered the seeds necessary to create and grow this program. Kathryn's big brothers **Arnold and Fred, and Clara, Maura, Jack, Sophie, Charles, and Sarah,** who continue to strengthen, build, and inspire Kathryn's Hope.

IN HONOR
In recognition of all in the world that struggled with Hopelessness in some way, shape or form, and left us way too early, including a few close to our hearts. Thank you for teaching us so much about life, love, and Hope. May we spread Hope far and wide in your name and honor:
Jon and Sally Goetzke, Tom Foorman, Dr. Stephen C. Gleason, Vicky Harrison, Eloise Land, Jesse Lewis, and **Austin Weirich.**

Recent statistics paint a startling picture of the current mental health landscape, particularly for our youth and minority communities. In the United States alone, 57% of teen girls are persistently Hopeless, and 30% are contemplating suicide. These figures are not isolated; 44% of all youth face persistent Hopelessness, with even higher rates in marginalized communities.

Hopelessness stands as a primary indicator across various mental health conditions. Globally, 1 in 7 10-19 year olds will experience a mental health disorder, and depression, anxiety, and behavioral disorders are the leading cause of illness and disability among adolescents. Hopelessness profoundly influences health outcomes, productivity, educational achievements, and human connections. Despite its pervasive impact, we are not taught how to recognize and actively manage Hopelessness. We must strategize; suicide is the 4th leading cause of death among 15-29-year-olds globally.

The distressing reality of Hopelessness and suicide extends beyond our youth. Over 700,000 lives are lost to suicide annually worldwide, which equates to 1 death every 40 seconds. A recent study from Harvard found that half of the world's population will grapple with a mental health disorder in their lifetime with anxiety, depression, posttraumatic stress disorder, alcohol misuse, and phobias affecting millions.

This public health crisis demands urgent attention. We do not have enough providers to address the health needs of the world and we are taking too long to take action. While hotlines and crisis interventions are essential, they merely scratch the surface. The prevalence of psychiatric emergencies entering our Emergency Rooms is growing. . We must dive deeper to table the root cause: Hopelessness. Waiting until individuals reach the emergency room before addressing their mental health struggles is no longer acceptable.

Our mission is clear: equip everyone with the tools to understand and nurture Hope. We want to empower communities with evidence-informed resources while building evidence-based programming for different sectors through ongoing research and partnerships. Hope is a well-researched critical strategy for preventing and mitigating suicide and mental health disorders.

Join us in activating this global movement to ensure all know the what, why, and how of Hope.

Dr. Myron Belfer, MD, MPA
Professor of Psychiatry at the Harvard Medical School and a Senior Associate in Psychiatry at Boston Children's Hospital

I write to you about a silent epidemic gripping our youth: Hopelessness. It's a phenomenon that holds dire consequences, predicting a myriad of risky behaviors among our youth. From violence to self-harm, and bullying to partner violence, Hopelessness serves as a consistent predictor of these distressing behaviors. Across the world, 1 in 3 women are victims of violence and 475,000 people are victims of homicide annually.

I am witnessing the repercussions of Hopelessness firsthand in the emergency room day in and day out; youth are coming in with gunshot wounds, stabbings, and trauma. I talk to parents, grandparents, brothers and sisters. We are a society in crisis.

Gun violence is predicted by a history of violence, and Hopelessness is the only consistent predictor of violence. Yet we continue treating people in the emergency room, doing our best to cover their wounds, instead of working collectively to prevent the wounds from occurring. What I witness is preventable, yet it takes collective action.

We know that Hopelessness is often a consequence of oppression and discrimination, so it's learned. We are teaching our kids to be Hopeless, and that continues the ongoing cycle of oppression. If we want to stop the cycle of violence, self-harm, addiction, and bullying, we must start ensuring all know what Hopelessness is and skills to activate its antidote: Hope.

Please join us to activate your city and community on these critical skills. We need to amplify the message that Hope is not a wish it is measurable and teachable. We must take Hope out of the abstract and give all the skills and tools they need to proactively manage Hopelessness and get to Hope. And the time for Hope is now.

Instead of punishing kids, and putting them in jails, we need to arm them with skills to navigate the many challenges they face. We need to elevate stories of individuals that have successfully navigated their specific challenges, and share specific strategies for how these Hope champions managed stress, practiced happiness habits, took inspired actions, cultivated nourishing networks, and eliminated negative thinking challenges.

We must equip them with skills for 'how' to Hope.

Join us to create a global movement for Hope.

Dr. Edward Barksdale, Jr., MD
Robert J. Izant, Jr. MD Professor and Surgeon-in-Chief at Rainbow Babies and Children's Hospital/ University Hospitals and Case Western Reserve School of Medicine (CWRUSOM)

Hopelessness is quietly corroding the fabric of our cities. It is fueling homelessness, addiction, anxiety, depression, accidents, dropouts, risky behaviors in our youth, and teen pregnancies. Yet, how many of us are truly knowledgeable about Hopelessness and how to nurture Hope?

Hopeful Cities® has a clear mission: to take a proactive stance in tackling Hopelessness. We envision a unified strategy for fostering Hope encompassing the whole city and person. It's about normalizing conversations around moments of Hopelessness and ensuring everyone possesses the skills to cultivate Hope.

The 2030 Agenda for Sustainable Development[1], adopted by all United Nations Member States in 2015, provides a shared blueprint for peace and prosperity for people and the planet, now and into the future. The July 2023 Sustainable Development Goals Report Special Edition[2] showed in the last three years, progress on more than 50 percent of targets of the SDGs is weak and insufficient; on 30 percent, it has stalled or gone into reverse.

Additionally, the report noted that under the current trajectory, by 2030, nearly 575 million people will be living in extreme poverty levels. Moreover, currently, nearly 1.6 billion people in the world are homeless or have inadequate housing. Additionally, the global temperature has risen 1.1 degrees Celsius since pre-industrial times and is quickly approaching the 1.5 degrees Celsius tipping point.

Failure to make progress towards goals[3] may lead to negative affective states, including anxiety and depression, which impact a city's functioning and health. Since the start of the COVID-19 pandemic, there has been a 13% rise in mental health conditions worldwide, and depression is the leading cause of disability.

Hope is a protective factor for anxiety and depression[4] and is teachable. Action is imperative, and we must act early. Hopeful Cities involves multiple sectors, including education, art, workplaces, science, government, and public health awareness campaigns. We can establish a universal language of Hope and equip all areas of the community with the ability to recognize Hopelessness, measure their own Hope, and acquire the necessary skills to enhance Hope.

Hopelessness is putting a strain on our cities' resources, so we can no longer delay enacting a strategy for Hope. Let's work together to build cities that empower and support, ensuring that Hope is not just a concept but a tangible reality for all.

Sincerely,

Mayor Hillary Schieve
Reno, Nevada

1. sdgs.un.org/2030agenda
2. unstats.un.org/sdgs/report/2023/The-Sustainable-Development-Goals-Report-2023.pdf
3. www.ncbi.nlm.nih.gov/pmc/articles/PMC3864849/
4. www.ncbi.nlm.nih.gov/pmc/articles/PMC9197088/

In this critical juncture of humanity, marked by wars, natural disasters, a climate crisis, stalled progress on Sustainable Development Goals (SDGs), and pandemics, we face the unprecedented challenge of the worst youth mental health crisis in history. Everyone across the world faces Hopelessness, yet our responses too often come only after crises have unfolded. Waiting until a crisis is also the most ineffective and expensive time to address an issue. We must and can do better.

It is evident that we need a sustainable solution to address the multifaceted outcomes of persistent Hopelessness, including addiction, school shootings, partner violence, suicide, and mental health issues. However, our tendency to focus on the problem rather than the solution perpetuates a cycle of delayed action. Stigma stops an alarming 50% of people with mental health problems from getting treatment in the US, a rate of up to 90% in many other countries, and we don't have enough mental health professionals and jail facilities to address the growing challenge of violence, addiction, anxiety, depression, and suicide.

Hope emerges as the transformative strategy to initiate a global movement for change, embracing a whole society approach. Hope bypasses stigma, as the skills to Hope can be taught universally to all people across the lifespan. By employing consistent branding and language, Hope becomes a beacon that transcends borders, reaching every corner of the world, even those living without adequate resources.

What sets Hope apart is its accessibility – a low-cost or no-cost strategy that makes it inherently suitable for widespread implementation. In our pursuit of global Hope activation, my inspiration arises from personal experiences. The tragic loss of my late father to suicide when I was only 18 and my own dive in to Hopelessness through a suicide attempt, addiction, self-harm, anxiety, depression, and PTSD. This year I celebrate 20 years of recovery, and ten years free of medications due to my consistent and dedicated approach to Hope using our Shine Hope framework. I won't say Hope is easy, yet it is always possible.

Recognizing that every individual encounters moments of Hopelessness, our mission is crystal clear: to remove the shame that often surrounds violence and mental health, and ensure that everyone understands what Hopelessness is, how to proactively manage it, and be equipped with skills to activate Hope. This is not a medicalized approach to mental health; rather, it is a cost-effective, proactive solution we can all practice together. Our aim is to empower individuals with the skills to self-manage, cultivate strong peers for support, and guide all to community support when additional help is needed.

As we embark on this transformative journey, I extend an invitation to join us in this global movement for Hope. Together, let us create a world where Hope is not just an obscure concept but a shared reality, allowing everyone to proactively manage their moments of Hopelessness and always finding their way back to Hope.

Kathryn Goetzke, MBA
CEO and Chief Hope Officer, The Shine Hope Company
Founder, iFred and Creator, Hopeful Cities

Once you choose Hope, anything's possible.

-Christopher Reeve

Table of Contents

3 Letter from Dr. Myron Belfer, MD, MPA
4 Letter from Dr. Edward Barksdale, Jr., MD
5 Letter from Mayor Hillary Schieve of Reno, NV
6 Letter from Kathryn Goetzke, MBA

10 Executive Summary
The Why
Hopelessness and Hope Stats
Our Research
Hope and Sustainable Development Goals

23 Section 1: Hope, Hopelessness, and the Brand
Hopelessness is Affecting your City
Hope: Empowering Cities
Power of a Brand
Creating a Global Movement for Hope

35 Section 2: Eight Guiding Principles
Take a Whole Community Approach
Bridge the Knowledge Action Gap
Utilize Solution-Focused Methods
Act Early Everywhere
Empower All Citizens
Amplify a Universal Brand
Use Evidence-Informed and Evidence-Based
Activate the Shine Framework

41 Section 3 Six Sectors for Fostering Hope
Government
Science
Education
Workplace
Healthcare
Art

92 Appendix
162 Links and References
172 Additional Resources
174 Special Acknowledgement from the Author

Executive Summary

The Why

Hopelessness, defined as emotional despair and motivational helplessness, is impacting our cities. Every single person experiences moments of Hopelessness, big or small, from something as simple as traffic or missing a bus to something big like failing a test, losing a job or loved one. Left unmanaged, these moments of Hopelessness can turn into persistent Hopelessness, and persistent Hopelessness is linked to many adverse outcomes that directly affect cities, such as economic decline, increased crime rate, addiction, and impacts the mental health and overall well-being of residents.

The antidote to Hopelessness is Hope, defined in our work as a vision for something in the future, fuled by both positive feelings and inspired actions. As we have shown, Hope is a teachable, measurable, and learnable skill; however, we are not taught how to Hope in school. Research has consistently demonstrated the power of Hope as it relates to improved productivity, retention, education outcomes, and mental health, safety, and health, while decreasing addiction, violence, and recidivism.

Hope is also critical to all Sustainable Development Goals (SDGs) set for by the United Nations. Individuals who are Hopeful set and achieve their goals while managing challenges and maintaining a positive mindset; these are crucial aspects of the goal attainment strategies needed to meet the SDGs. Further, lack of goal attainment can lead to clinical anxiety and depression, and based on recent reports we must ensure all working on SDGs are equipped with skills to proactively manage these challenges and activate Hope.

Through this Hopeful Cities Playbook, we aim to give cities tools to ignite a transformative movement centered on activating Hope within every individual and community. We are empowering cities with Hope programming to amplify Hope science in the following six sectors; **government, science, healthcare, education, art,** and the **workplace.** We have eight guiding principles that are in all programming to drive innovation and messaging, that include:

By integrating Hope strategies into every facet of life, we envision a world where individuals harness their inner strength to overcome adversity, creating vibrant, connected communities. We are committed to providing accessible resources, fostering collaboration, and driving initiatives that elevate mental wellness, ultimately cultivating a more Hopeful and thriving global landscape for generations to come.

HOPELESSNESS IS A KEY PREDICTOR OF THE FOLLOWING OUTCOMES:

VIOLENCE

4.4 MILLION

People killed each year globally.

726 MILLION WOMEN GLOBALLY (or nearly 1 in three) have been a victim of physical or sexual partner violence at least once in their life.

GENDER INEQUALITY

286 years to close gender gaps in legal protection and eliminate discriminatory laws. *(based on current trajectory)*

CLIMATE CRISIS

1.1°C Global Temperature Rise since pre-industrial levels

projected to exceed the 1.5°C tipping point by 2035 subsequently increasing heat waves, droughts, floods, wildfires, and rising sea levels.

DISPLACEMENT

110 MILLION PEOPLE DISPLACED

due to conflicts and human rights violations.

with 35 MILLION BEING REFUGEES

–the highest recorded figures.

WORKFORCE COSTS

Each year, employers face a cost of

$15,000 PER EMPLOYEE WITH DEPRESSION

due to decreased productivity, healthcare expenses, and turnover, **the greatest cost to the workplace.**

11.5 DAYS Diminished productivity every three months *(half of every eight hour shift.)*

ADDICTION

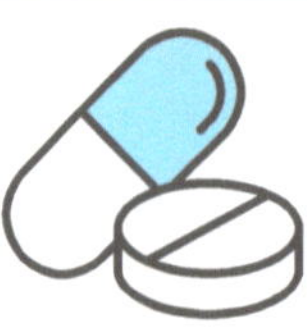

26% rise of **substance use** across the world.

Hopeless individuals are starting to use substances at an earlier age.

MENTAL HEALTH

Annual Cost of Depression Worldwide

US $1 TRILLION

Poor mental health cost the world economy

$2-5 TRILLION per year

Projected to rise to **$6 Trillion by 2030.**

POVERTY

Pandemic reversed 30 years of poverty reduction progress.

Global struggles with surging debt, inflation, trade tensions, and limited fiscal capacity.

2030

FORECAST:

575 MILLION PEOPLE

living in extreme poverty.

INADEQUATE EDUCATION

84 MILLION CHILDREN OUT OF SCHOOL BY 2030

Underinvestment and learning losses forecast

300 Million children or young people currently attending school will leave without basic literacy skills.

POOR HEALTH

- Predictor for unintentional injury and chronic health conditions
- impacts self-care during illness
- increases the risk of comorbid conditions
- contributing to poorer prognosis in conditions

HOPELESSNESS IS LEARNED.

HOPE IS A STRATEGY TO PREVENT AND REDUCE

VIOLENCE

HOPEFULNESS is linked to lower levels of violence

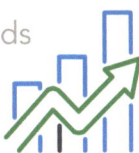

ADDICTION

Adolescents who are more Hopeful are less likely to use substances

GENDER INEQUALITY

HOPE emerges as a powerful construct for support and pursuing social change

MENTAL HEALTH

Hope protects against mental health disorders and can act as an intervention for existing mental health concerns.

POVERTY

People with Hope invest more into their future, which can lead to better financial outcomes.

CLIMATE CRISIS

HOPE drives group motivation toward outcomes beyond an individual's control, **a key factor in addressing climate change.**

INADEQUATE EDUCATION

HIGHER HOPE INCREASED SCHOOL RETENTION BY 53%

Increased graduation rate even in areas where 1 in 3 kids graduate due to community hardship and adversity.

85% of K-12 school superintendents report student Hopefulness is a "very important" way to measure the effectiveness of public schools

DISPLACEMENT

HOPE PLAYS A VITAL ROLE IN CONFLICT RESOLUTION

serving as an intervention that fosters peace in resolving conflicts, which may help reduce the number of refugees, and Hope also helps refugees manage change

POOR HEALTH

Hope enhances physical health and outcomes, lowering the risk of mortality and chronic conditions

(e.g., diabetes, hypertension, stroke, cancer, heart and lung diseases, arthritis, obesity, and chronic pain), while also boosting treatment adherence and speeding up recovery.

WORKFORCE COSTS

Hope is one of the four basic needs in Strength-Based Leadership

14% INCREASE

in workplace productivity, outperforming intelligence, optimism, and self-efficacy.

Higher Hope is also linked to increased engagement and employee retention

HOPE IS TEACHABLE.

*See links and references on Page 158

Our Research

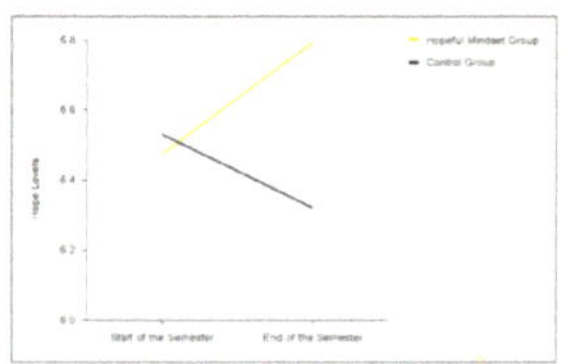

Our Founder has been working on mental health strategy the past 20 years, with a focus on Hope the past ten. She has been studying Hope globally for the last 10 years, conducting focus groups, doing interviews, conducting secondary research, and conducting primary research. We aim to continue improving our research with your help on our programs.

In our research endeavors to date, Kirby et al. (2021) explored the efficacy of the "Hopeful Minds" curriculum, revealing substantial enhancements in various aspects of well-being among children. Specifically, improvements were observed in anxiety, depression, resilience, emotional regulation, social support seeking, and self-care.

Furthermore, Ghazali et al. (2021) conducted a study focusing on primary school children in Malaysia, demonstrating that Hope education significantly decreased depression symptoms and enhanced negative emotion regulation.

Another study by Kirby et al. (2021) emphasized the positive impact of the "Hopeful Minds" program, showcasing increased emotional insight, resilience, confidence, and the development of new coping skills among 8-14-year-olds.

Throughout these studies, we can conclude ***Hope is Teachable.***

Hope and Sustainable Development Goals

Researchers have found that individuals with higher levels of Hope are more likely to achieve their goals, subsequently improving their well-being (Moss, 2018). The goals are met because these individuals have high agency-related Hope thoughts (i.e., the belief that they can attain their goals and are successful in life) and pathways-related Hope thoughts (i.e., the belief that they can overcome barriers and develop alternative solutions to goals when needed (Oettingen & Gollwitzer, 2002).

These goals are all vital for the health and vitality of our cities, and our world.

As individuals with higher levels of Hope are more likely to achieve their goals, ***individuals high in Hope are critical for reaching all SDGs***, not just a target under a goal, as the case was made for mental health.

These SDGs recognize that ending poverty and other deprivations must go hand-in-hand with strategies that improve health and education, reduce inequality, and spur economic growth – all while tackling climate change and working to preserve our oceans and forests. Hopelessness is learned and is fueling our lack of progress toward the SDGs; thus, at the foundation of each of these is the need for activated Hope.

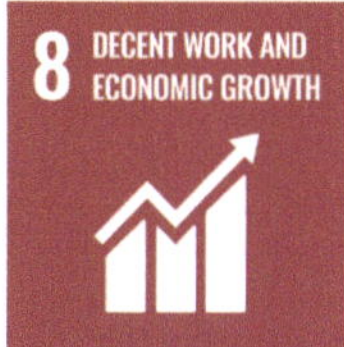

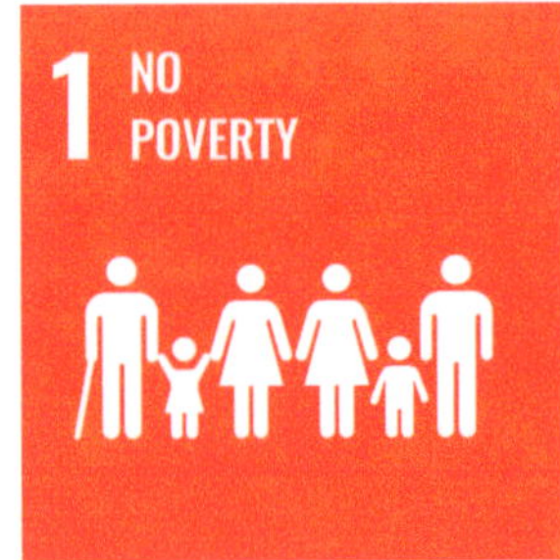

Goal 1: End poverty in all its forms everywhere.

Individuals living in impoverished conditions are prone to higher levels of Hopelessness, while those experiencing elevated Hopelessness are more likely to endure increased poverty (Patel & Kleinman, 2003; Ridley et al., 2020). The entanglement of Hopelessness and stress contributes to a poverty trap, anchoring individuals at the poverty level. In contrast, individuals with a positive outlook on their future, characterized by higher levels of Hope, tend to make more substantial investments in their future, resulting in improved financial outcomes (Eggers et al., 2003).

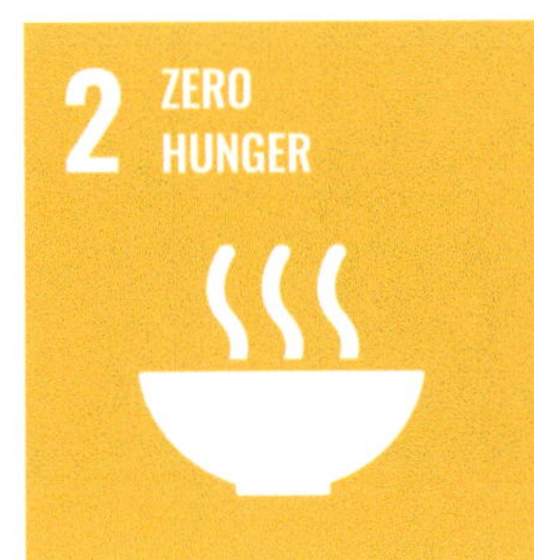

Goal 2: End hunger, achieve food security and improved nutrition, and promote sustainable agriculture.

Researchers found that higher levels of Hope strengthen the relationship between financial capabilities and one's perception of access to resources (Gilbert & Ashley, 2020). Moreover, Hope is also linked to agricultural technology adoption; individuals with elevated levels of Hope are more inclined to take proactive steps toward achieving their goals, making Hope a catalyst for embracing innovative farming techniques and improving our food supply (Bukchin and Kerret, 2018).

Alarming data from the Centers for Disease Prevention and Control underscored the pressing issue of farmer suicides, with farmers being twice as likely as individuals in other occupations to die by suicide (Peterson et al., 2020). Farming is one of the most perilous industries; thus, it is imperative to acknowledge Hope's role in buffering against suicide (Behere & Bhise, 2009; Huen et al., 2015).

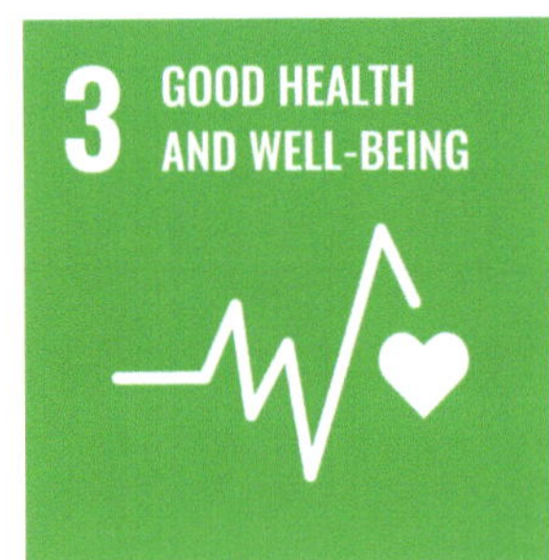

Goal 3: Ensure healthy lives and promote well-being for all at all ages.

Hope plays a crucial role in promoting health and well-being across various aspects of life, potentially preventing numerous health conditions. This is because individuals who are higher in Hope have an optimistic view of life leading them to take action to preserve their health (Harvard T.H. Chan, 2021). For example, individuals with elevated Hope often adopt healthier behaviors, such as more frequent exercise, reduced fat intake, and avoidance of substance misuse (e.g., Berg et al., 2011; Meraz et al., 2023; Nsamenan & Hirsch, 2014).

These positive health behaviors, as demonstrated by Harvard University's "The Human Flourishing Program," are linked to improved physical health and health-related outcomes, including a reduced risk of all-cause mortality, fewer chronic conditions (i.e., diabetes, hypertension, stroke, cancer, heart disease, lung disease, arthritis, and overweight/obesity, chronic pain), reduced risk of some mental health conditions (i.e., depression, anxiety, and stress), and improved sleep patterns (Feldman & Sills, 2013; Long et al., 2020; Senger, 2023).

Additionally, we know that Hope is related to an increased sense of connectedness and belonging, which is also linked to prevention and positive health outcomes (Wothington, 2020). In addition to prevention efforts, Hope is a documented intervention in health outcomes. For example, researchers have found individuals with higher Hope adhere to treatment plans better than those low in Hope, as they are more motivated to reach recovery goals (e.g., Javanmardifard et al., 2020; Kurita et al., 2020).

Moreover, research indicates that higher Hope is linked to expedited recovery times from injuries and diseases, with individuals higher in Hope having a more favorable prognosis for postoperative recovery (Long et al., 2020; Zhu et al., 2017; Zou et al., 2022).

Goal 4: Ensure inclusive and equitable quality education and promote lifelong learning opportunities for all.

Hope emerges as a crucial factor that predicts academic success and plays a pivotal role in fostering a positive and inclusive learning environment. (e.g., Bryce et al., 2019; Day et al., 2010; Dixson & Stevens, 2018) Researchers found that students high in Hope also have stronger friendships, more creativity, and better problem-solving, which are all skills to help them navigate the challenges of school and work toward graduation (Zakrzewski, 2012). Additionally, Hope helps protect against depression and anxiety, which are predictors of school dropout (Hjorth et al., 2016).

Hope also predicts academic achievement better than intellectual functioning (Bryce et al., 2019; Dixson & Stevens, 2018), and studies consistently show that students with higher levels of Hope tend to invest more effort into their educational goals and persist in the face of challenges (Idan & Margalit, 2013) Halpin (2001) also reported that teachers higher in Hope were able to motivate the most disaffected students. Students with higher Hope levels during their first year of college consistently demonstrate significantly higher grades at the end of three years, indicating that Hope is not merely a short-term motivator but a sustained driver of success (Day et al., 2010). Moreover, the significance of Hope is particularly evident in the context of students with learning disabilities who often face additional hurdles (Idan & Margalit, 2013).

Goal 5: Achieve gender equality and empower all women and girls.

According to the CDC (2023), in the United States, the prevalence of Hopelessness has doubled for girls compared to boys.

Globally, women and girls encounter discrimination based on sex and gender, contributing to various challenges such as domestic and sexual violence, unequal pay, limited access to education, and insufficient healthcare (Amnesty International, 2023). Recognizing the profound impact of Hopelessness on health, education, and workplace outcomes, it becomes imperative to equip women and girls will the skills to cultivate Hope as a strategy for promoting gender equality.

Goal 6: Ensure availability and sustainable management of water and sanitation for all.

The adverse effects of climate change-induced challenges (e.g., flooding, tornados, etc.) pose physical threats and people exposed to such disasters often suffer from acute traumatic stress, and their symptoms persist until safety and security are reestablished (Fritze et al., 2008). Furthermore, the looming threat of essential resources, such as water, becoming more scarce due to climate change intensifies anxiety among populations (Fritze et al., 2008). Increasing Hope protects it against the anxiety related to water scarcity while also opening up pathways for innovative solutions and collaborative efforts to address the pressing problems of water scarcity (Fritze et al., 2008; Rahimipour et al., 2015).

Goal 7: Ensure access to affordable, reliable, sustainable, and modern energy for all.

Individuals with higher levels of Hope exhibit a greater capacity to overcome obstacles and discover innovative pathways toward the goal of achieving clean energy (Marlon et al., 2019). Moreover, elevated levels of Hope contribute to enhanced problem-solving abilities, particularly in addressing the multifaceted challenges associated with affordable, reliable, and sustainable energy solutions (Chang, 1998; Geiger et al., 2019). Research underscores the significance of constructive Hope, defined as making progress on clean energy, coupled with constructive doubt, acknowledging the reality of the threat of climate change (Marlon et al., 2019).

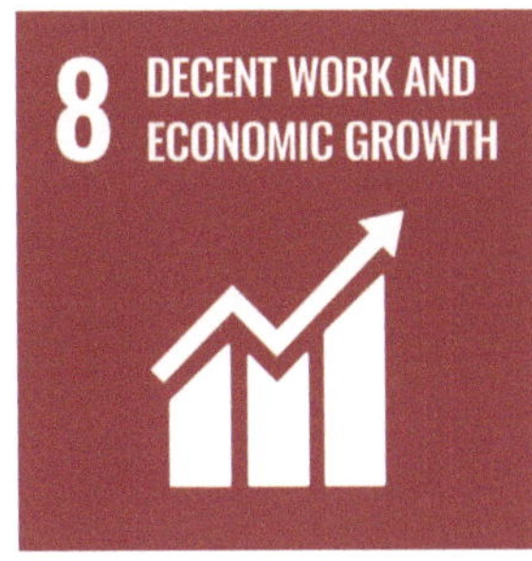

Goal 8: Promote sustained, inclusive, and sustainable economic growth, full and productive employment and decent work for all.

The significance of Hope in the workplace is underscored by a Gallup study, which identified Hope as one of the four primary needs of employees (Berg, 2020). The Gallup study's measure of Hope in the workplace revealed that employees who strongly agree that their leaders make them feel enthusiastic about the future are 69 times more likely to be engaged in their work compared to those who disagree with this statement (Berg, 2020).

Moreover, the impact of Hope on productivity is substantial, accounting for about 14% of work productivity, surpassing the influences of intelligence, optimism, or self-efficacy (Lopez, 2013). Additionally, Hope serves as a protective factor against anxiety and depression, mitigating the economic burden associated with lost productivity (Hicks & McFarland, 2020). Depression and anxiety alone are estimated to cost the global economy USD 1 trillion per year, with projections reaching USD 16 trillion by 2030 (Chodavadia et al., 2023).

Goal 9: Build resilient infrastructure, promote inclusive and sustainable industrialization, and foster innovation.

Individuals with higher levels of Hope exhibit superior problem-solving skills and enhanced creativity, essential elements for the development of resilient infrastructure and the promotion of inclusive and sustainable industrialization while fostering innovation. Hope produces an optimistic mindset, allowing individuals to approach the challenges of building infrastructure and industrialization as opportunities for growth, leading to effective problem-solving and the generation of innovative solutions (Weronika et al., 2022).

Additionally, Hope serves as a precursor to resilience, instilling the confidence and determination needed to overcome setbacks and persist in the face of challenges (Senger, 2023).

Goal 10: Reduce income inequality within and among countries.

Hopelessness is often a consequence of oppression and discrimination, so it continues the cycle of oppression as all the related negative outcomes of Hopelessness are in the cycle of poverty (Mitchell et al., 2020).

Teaching Hope, especially in disadvantaged communities, is a strategy to improve self-efficacy to create strategies to get out of poverty both within and among countries.

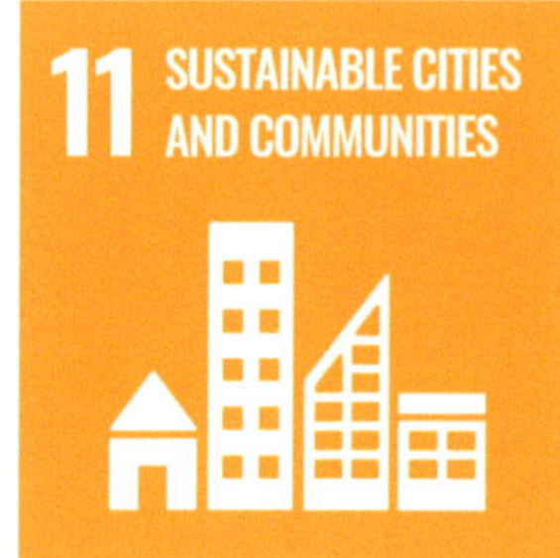

Goal 11: Make cities and human settlements inclusive, safe, resilient, and sustainable.

Hope plays a pivotal role in enhancing the safety and resilience of cities. Research demonstrates that elevated levels of Hope contribute to a reduction in substance use and criminal behaviors (Brooks et al., 2016; Martin & Stermac, 2010; Mathis et al., 2009). Furthermore, a positive correlation exists between higher levels of Hope and a decreased likelihood of recidivism, leading to a subsequent reduction in overall crime rates and improving the safety of cities and communities (Dekhtyar et al., 2012). Additionally, Hope is associated with a decline in sexual risk behaviors and teen violence, as indicated by various studies (Hill et al., 2019; Li et al., 2022).

Goal 12: Ensure sustainable consumption and production patterns.

Extensive research suggests that personal motivation, environmental motivation, and Hope are robust predictors of early adoption of environmentally friendly technologies (Bukchin & Kerret, 2019). In this context, Hope plays a pivotal role in encouraging responsible energy consumption, as higher levels of Hope are linked to a greater propensity to embrace greener alternatives, thereby contributing to sustainable consumption practices. Furthermore, by increasing collective Hope within communities, there can be widespread shifts towards sustainable consumption and production patterns.

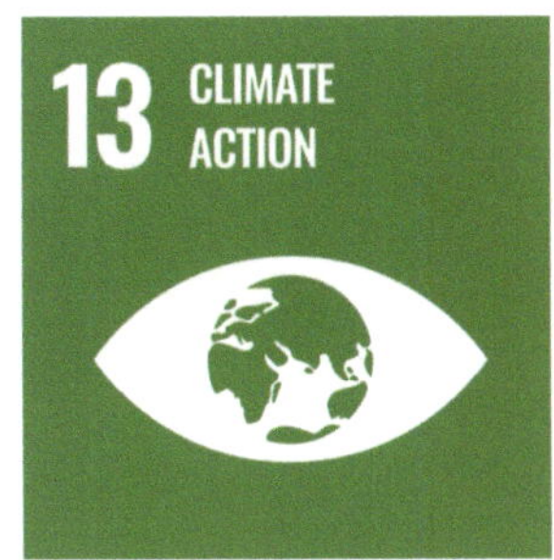

Goal 13: Take urgent action to combat climate change and its impacts by regulating emissions and promoting developments in renewable energy.

Numerous studies have underscored the positive relationship between Hope and engagement in climate-change-related activities, as well as pro-environmental behaviors (e.g., Bury et al., 2019; Ojala, 2023; Ojala, 2011). For example, (Bury et al., 2019) found that Hope can be used as an antecedent for group motivation toward an outcome that is not within one's sole ability to obtain, which is applicable to combatting climate change. Moreover, (Sangervo et al., 2022) found that a combination of climate-change-related anxiety and Hope were adaptive emotions and could be used as a motivator for taking action against climate change.

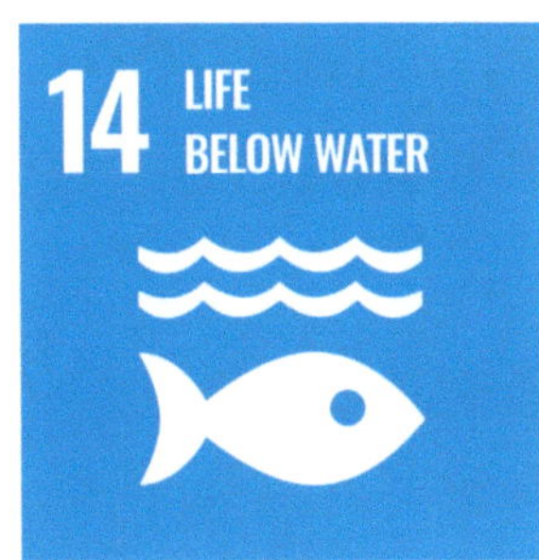

Goal 14: Conserve and sustainably use the oceans, seas and marine resources for sustainable development.

There is a two-way relationship between ocean health and human well-being, as the ocean accounts for 20% of our food supply (Costello et al., 2020; NOAA, 2023). Moreover, how we manage and take care of the ocean directly impacts our health, as pollution affects our food supply and leads to ocean-related natural disasters (i.e., tsunamis and hurricanes; NOAA, 2020).

Hope is a strategy to help sustain our ocean's health, as individuals with higher Hope are better at problem-solving, and we need effective strategies to prevent and reduce ocean pollution (Zakrzewski, 2012).

Goal 15: Protect, restore and promote sustainable use of terrestrial ecosystems, sustainably manage forests, combat desertification, and halt and reverse land degradation and halt biodiversity loss.

Hope demonstrates a positive correlation with pro-environmental actions, encompassing the sustainable utilization of land and the incorporation of green technologies in agriculture (Bukshin & Kerret, 2020; Peles & Kerret, 2021). Consequently, when individuals harbor Hope, farmers are more inclined to embrace environmentally friendly agricultural technologies, thereby reducing land degradation resulting from pesticides or suboptimal crop yields (Wang, 2022).

Goal 16: Promote peaceful and inclusive societies for sustainable development, provide access to justice for all and build effective, accountable and inclusive institutions at all levels.

Hopelessness predicts all risky behaviors in youth (i.e., violence, weapon-carrying on school property, self-harm, unprotected sex, bullying, partner violence, etc.; Bolland, 2003). Moreover, the only known predictor of gun violence is a history of violent behavior, and researchers have found Hopelessness is the only consistent predictor of violent behaviors (APA, 2013; Demetropoulos, 2017). However, Hope is an important construct for conflict resolution; Hope is an intervention that can promote peace when resolving intractable conflicts (Cohen-Chen et al., 2013).

Goal 17: Strengthen the means of implementation and revitalize the global partnership for sustainable development.

Hope serves as a documented effective strategy for building collaboration amongst groups of people. Researchers have found individuals with higher levels of Hope have greater positive connections with others, are more collaborative, and are better problem-solvers, which is necessary for strengthening the means of implementation and cultivation of healthy partnerships for sustainable development (Merolla et al., 2021).

SECTION 1

Hope, Hopelessness, and the Brand

Hopelessness is Affecting Your Community

Embracing a city-wide approach to Hope is pivotal because everyone encounters moments of Hopelessness. **Every. Single. Citizen.** How so?

HOPELESSNESS is characterized by
EMOTIONAL DESPAIR *(sadness, anger, fear)* and
MOTIVATIONAL HELPLESSNESS *(a sense of powerlessness)*

We all experience these moments. They can surface in seemingly minor situations like being stuck in traffic, facing job rejections, or dealing with technology glitches. Yet, these seemingly small instances can spiral into larger issues when compounded by events such as job loss, financial turmoil, or relationship breakdowns.

On a grander scale, systemic oppression, violence, and global crises, like climate change, also contribute to a collective, persistent sense of Hopelessness, which can impact the overall functioning and safety of an individual and a city.

Hopelessness impacts everyone, so rather than eradicating Hopelessness, we must teach all 'how' to manage these moments of Hopelessness to prevent persistent Hopelessness. We must normalize them, instead of shaming people for experiencing them, and ensuring all know how and where to get support.

The consequences of unmanaged Hopelessness are far-reaching and alarming; Hopelessness is linked to:

- **Increased recidivism:** **More than 50% of people released from prison return within a year;** Hopelessness leads to substance use, lack of desire to treat mental health problems, and interpersonal conflicts, which ultimately leads to recidivism (Benecchi, 2021; Dodd, 2016).

- **Increased violence:** The only known predictor of gun violence is a history of violent behavior, and researchers have found Hopelessness is the only consistent predictor of violent behaviors (APA, 2013; Demetropoulos, 2017). In 2019, there were 250,227 gun-related deaths worldwide (World Population Review, 2019).

Increased susceptibility to mental health disorders: Anxiety and depression are connected to feelings of Hopelessness, with depression standing as the primary cause of disability globally, **which results in an annual cost of $1 trillion to the global economy** (Chodavadia et al., 2023; WHO, 2023).

Increased risk of homelessness: Nearly 154 million people are homeless, and depression is a precursor of homelessness for both men and women. (Kuo, 2019; Moschion & Ours, 2022).

Increased suicide: Across the world, suicide is one of the leading causes of death (WHO, 2021).

Increased health problems: Health-related outcomes related to Hopelessness include, unintentional injury, chronic conditions (e.g., cancer, cardiovascular disease, hypertension), higher risk of comorbid conditions, poorer prognosis, and chronic pain (de Faria et al., 2020; Everson et al., 1996; Katon, 2011)

Increased Substance Use: The world has seen a 26% rise in substance use in the past decade, and Hopelessness is directly linked to the use of all substances (Kuo et al., 2004; UNODC, 2022)

Traditionally, society tends to react only when crises are at their peak, neglecting the early signs. However, a shift to a proactive approach is imperative. Equipping everyone with an understanding of Hopelessness enables them to recognize its signs and acquire skills to pivot toward Hope. This proactive stance is a potent preventive measure against the larger negative repercussions associated with unmanaged Hopelessness.

Our strategy, encapsulated in the Hopeful Cities Playbook, adopts a holistic approach. It aims to educate and empower every individual within a city using a shared language centered around Hope. Leaders and citizens will work together to learn the same skills and support each other in the process.

The focus is not to label individuals, but to impart Hope as a skill. Much like seeking assistance for tutoring or coaching, cities can seek guidance and support in navigating the challenges linked to moments of Hopelessness. Mayors are equipped with skills to be Heroes of Hope.

Hope: Empowering Cities

Contrary to popular belief, Hope isn't just wishful thinking:

Dr. Myron Belfer, a Hope expert, describes ***Hope as something we have control over. It's a skill and a motivation, so it is something that we can work towards.***

Dr. Dan Tamasulo says that ***Hope is a reorganization of perceptions to foster the belief that you have control in the future.***

Dr. Shane Lopez explains that ***Hope is the feeling you have when you have a goal, are excited about achieving your goal, and then figure out how to achieve your goal.***

We have combined a number of these principles into the Hopeful Cities Playbook and have developed the following definition using Hopelessness as the antithesis:

HOPE is a vision for something in your future, fueled by both **POSITIVE FEELINGS** and **INSPIRED ACTIONS.**

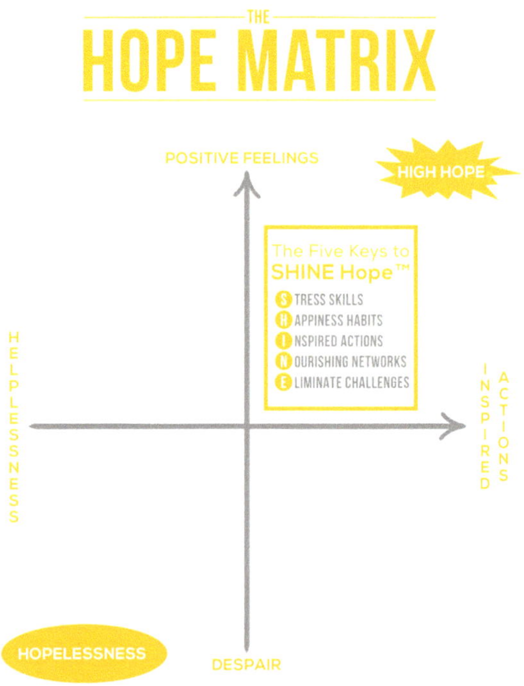

We've created The Hope Matrix visual to make it easier to understand. Any challenge in life can be navigated using Shine Hope™ skills to move from Hopelessness to Hope. Why does it matter?

Researchers have found that individuals with higher levels of Hope are more likely to achieve their goals, subsequently improving their well-being (Moss, 2018). The goals are met because these individuals have high agency-related Hope thoughts (i.e., the belief that they can attain their goals and are successful in life) and pathways-related Hope thoughts (i.e., the belief that they can overcome barriers and develop alternative solutions to goals when needed (Oettingen & Gollwitzer, 2002).

The benefits of nurturing Hope are far-reaching and transformative. Research consistently underscores that people higher in Hope have numerous positive outcomes, such as -

IMPROVED EDUCATION

Hope drives significant improvements in education, surpassing intelligence and prior academic performance, predicting a GPA boost of 0.21 points per additional point in Hope score, and fostering higher graduation rates, especially in communities facing considerable hardship and adversity (Bashant, 2016; Curry et al., 1997; Day et al., 2010; Gallagher et al., 2017).

IMPROVED WORK PERFORMANCE

By reducing anxiety and depression while significantly boosting workplace productivity by 14%—outperforming measures of intelligence, optimism, and self-efficacy—Hope plays a pivotal role in enhancing employee retention, job engagement, sleep quality, and expedited recovery times after injuries, ultimately mitigating substantial financial losses for employers due to Hopelessness (APA, 2013a; APA, 2013b; Baraket-Bojmel et al., 2023; eHansan et al., 2022; Trezise et al., 2018; Warwick, 2012).

IMPROVED HEALTH

Through reducing risky behaviors, decreasing the risk of chronic health conditions, improving medication adherence to curb healthcare costs, expediting recovery post-injury, fostering resilience against stress, and alleviating depression and anxiety—leading causes of disability globally—Hope significantly contributes to overall health (Brooks et al., 2016; Harvard T.H. Chan, 2021; Hill et al., 2018; Morgado & Cerqueira, 2018; Ong et al., 2018; Warwick, 2012).

IMPROVED SOCIAL CONNECTIONS

Individuals with higher levels of Hope have greater positive connections with others, are more collaborative, and are better problem-solvers, which is necessary for strengthening the means of implementation and cultivation of a healthy community (Merolla et al., 2021).

IMPROVED SDG ATTAINMENT

Hope-driven initiatives prioritize well-being, promoting mental health and social support systems essential for achieving multiple SDGs. Moreover, this approach encourages long-term vision and planning, aligning city policies with the global sustainability agenda. Learn about specific SDGs and Hope here. theshinehopecompany.com/Hope-science/sustainable-development-goals

There is a robust scientific field studying Hope, and all of the research keeps telling us the same thing: Hope is important. You can see the extensive literature about the power of Hope on our website: theshinehopecompany.com/research. However, we are not born with Hope; it is a skill that must be developed and practiced, and we all have different starting levels of Hope.

Harnessing Hope as a strategy can revolutionize cities. By fostering a culture that understands, cultivates, and practices Hope together, communities can undergo numerous positive changes. Hopeful indivdiuals contribute to a more vibrant workorce, boosting productivity and reducing absenteeism. Effective leadership that nurtures Hope fosters trust, stability, and prolonged employee retention, which can result in a thriving economy.

Moreover, when cities prioritize Hope, they invest in the mental and emotional well-being of their residents. This proactive effort can prevent adverse outcomes linked to Hopelessness, such as crime and health-related challnges; it also helps foster a sense of community cohesion and shared purpose.

In essence, embracing Hope as a strategic pillar for city improvement isn't just about individual empowerment –it's about shaping resilient, thriving commnities where positivity and purpose converge to fuel progress and well-being.

"Hope impacts everything. How long you live, how quickly you recover from disease, your workplace performance and productivity, and your likelihood of graduation. Investing in Hope makes business sense."

- Kathryn Goetzke, MBA

Power of a Brand

Iconic symbols define brands and evoke powerful emotions, and the concept of a universal symbol for Hope can help build cohesion amongst a community during this city-wide initiative. Consider the resonance of symbols like the golden arches of McDonald's, the Nike swoosh, or the sleek Apple logo. They transcend mere visuals, imprinting lasting impressions that become synonymous with their respective entities, and they bring a feeling and attributes related to the iconic image.

What about 'Stop, Drop, and Roll'. Or 'Got Milk'? Or the infamous 'This is your brain. This is your brain on drugs'. While these public campaigns had powerful messages that everyone remembered, did they have impact? Did they elicit the change we desired? Some maybe yes, some maybe no. Either way, you remembered.

We've put a lot of energy into researching and understanding what works. Our aim is to create a universal symbol and brand, and public health campaign and strategy around Hope. Something that is bold, powerful, empowering, and unifying.

We want all to be inspired to Shine Hope, and embrace the skills we teach in the programming. We don't want to just talk about Hope, we want all know how to get to Hope. We are rebranding the word itself from just a 'wish', to a powerful force that transforms lives and even humanity.

Enter the sunflower– the global symbol for Hope. The sunflower is an iconic image that is grown around the world, symbolizing our ability to grow from darkness and reach towards the light. It is a universal representation of our ability to Shine Hope, as sunflowers are programmed to follow the light, even in challenging circumstances. They are grown in every country around the world, making them truly universal.

Example of our Company Logo and programs incorporating sunflower.

Brands hold immense sway over perceptions and emotions. They create indelible associations so that people remember the idea and internalize the message. McDonald's conjures thoughts of fast food, Nike elicits a sense of athleticism, and Apple evokes innovation. The repetition of these symbols ingrains them in our minds, wielding significant influence. We aim to replicate this influence with sunflowers and Shine Hope messaging (Stress Skills, Happiness Habits, Inspired Actions, Nourishing Networks, and Eliminating Challenges) so that when all see the sunflower and hear 'Shine Hope,' they know how to use the Shine Hope framework to achieve whatever they want in life.

And if they can't figure it out, they know where to go for support. And they know what to study from their heroes, so they can start modeling healthy behaviors. We want all, no matter what their circumstances, to be equipped to find their way from Hopelessness to Hope.

Why Sunflowers?

We were very deliberate in our choosing of the sunflowers. It all began from a study by the Emotional Impact of Flowers Study conducted by Jeannette M. Haviland-Jones, Ph.D., Professor of Psychology, Project Director, Human Development Lab at Rutgers. According to her research, regardless of age, flowers have an immediate impact on happiness. Recent studies have suggested flowers help reduce stress, and often increase serotonin and dopamine.

The evidence became clear as sunflowers were researched, as:

- The symbolism of the sunflower holds profound meaning. A sunflower seed begins its journey in darkness, mirroring our most Hopeless states. It represents our potential for growth and improvement amid despair. Just as a seed cannot flourish alone, we, too, rely on our Hope Network to nurture our Hope.

- The growth of a sunflower echoes our journey toward Hope. It stretches roots deep into the ground, akin to our efforts to break free from despair using **Stress Skills**—meditation, deep breathing, and mindful pauses.

- As the sunflower emerges into the sunlight, it unfurls leaves to gather sunshine, needing water, nourishment, and care to flourish. Similarly, we cultivate positive feelings through **Happiness Habits**—long-term, healthy practices fostering more and more Hope.

- Obstacles pepper the sunflower's path; rocky soil, and inadequate resources. Likewise, we face challenges. However, equipped with **Stress Skills**, **Happiness Habits**, **Inspired Actions**, **Nourishing Networks** and skills to **Eliminate Challenges**, we navigate and conquer these hurdles.

- The sunflower's purpose transcends its growth; it provides sustenance and joy. Similarly, we share Hope with those around us, becoming beacons of optimism and joy.

- Our choice of the sunflower and its vibrant yellow hue isn't arbitrary. It symbolizes our commitment to shine a positive light on Hope, eradicating mental health stigma through proactive measures in prevention, research, and education.

- In this endeavor, the sunflower becomes more than a symbol—it becomes the embodiment of Hope, illuminating pathways toward a brighter future for cities and individuals alike.

- It is also a method for nonprofits to raise funds for Hope. You can sell the seeds, have gardens sponsored, sell products in retail, or create art for auctions. The ideas are endless!

So, in this endeavor, the sunflower becomes more than a symbol—it becomes the embodiment of Hope, illuminating pathways toward a brighter future for cities and individuals alike.

"The sunflowers symbolize hope, and we're thrilled to expand Gardens for Hope, equipping children, teachers, and parents with tools for 'how' to hope."

- Ciara Byrne, Co-Founder & Co-CEO, Green Our Planet

SUNFLOWERS...

Fascinating Facts

The tallest sunflower was grown in the Netherlands by M. Heijmf in 1986 and was 25' 5.5" tall.

The sunflower plant reaches heights of 3 to 18 feet within just 90 to 100 days and produces a seed head with up to 2,000 seeds arranged in a spiral pattern.

The sunflower has heliotropic behaviors, meaning always turns their face toward the sun, which is a great metaphor for looking toward the future with hope.

I AM MANY

The yellow petals are actually the the protective leaves that cover the head while it is growing. That brown center is made up of hundreds of individually growing flowers, where a sunflower seed will emerge.

I LIKE EVERYBODY

Within the last few years, a new form of low-pollen sunflower has been created to help reduce the risk of asthma for pollen sufferers.

I HAVE POTENTIAL

Sunflower oil can be made into plastics, and research has also revealed that it the potential to create fuel for automobiles and other machinery.

Sunflowers are resilient and able to grow in harsh environments, such as dry and rocky soil. They grow out of the darkness and into the light, moving through challenging times with not much light. Yet they make it. This ability to thrive in tough conditions is a powerful symbol of hope and perseverance.

I AM ANCIENT!

The sunflower was domesticated from wild sunflowers around 1000 B.C. by Native Americans. It was ground into flours for making breads and soups, while sunflower oil softened leather, salved wounds, and conditioned hair.

I HELP THE ENVIRONMENT

An emerging technology called rhizofiltration, hydroponically grown sunflowers are grown floating over water. Their extensive root systems reach deep into sources of polluted water and extract large amounts of toxic metals, including uranium. The roots were able to extract 95% of the radioactivity in the water left behind by the accident at Chernobyl.

I BRING THE WORLD TOGETHER

Brought to the U.S. in 1996 from Russia, one of the largest areas of sunflower growth in the world, is in the former Soviet Union, which is second only to Argentina. The simple prairie sunflower, native to North America, is now one of the worlds leading oil seed crops behind soy beans.

Sunflowers are often used to symbolize new beginnings, growth, and renewal. This makes them a perfect symbol of hope, as they represent the idea of starting fresh and growing toward a brighter future.

"Keep your face to the sunshine and you cannot see the shadow. It's what sunflowers do."

Helen Keller

The Bonsai technique was used to make the shortest mature sunflower record. The sunflower was just over 2" tall.

I AM ROYAL

In Peru, the Aztecs worshipped sunflowers. They did so by placing sunflower images made of gold in their temples and crowning princesses in the bright yellow flowers.

Sunflowers are a bright yellow, and research suggests people who were surrounded by brighter and more vibrant colors reported higher levels of positive emotions, such as happiness and excitement.

Follow us: @ifredorg

iFred, a 501(c)3 organization, is working to teach hope, shine a positive light on mental health, and end the stigma around mental illness. iFred has started a number of programs to activate hope around the world, including Hopeful Minds®, Hopeful Cities®, and International Day of Hope. Learn more at www.ifred.org.

Fascinating Facts About Sunflower

PLANT SUNFLOWER GARDENS TO SHINE HOPE

Gardening is a great time to practice the Shine Hope Framework, as we have a lot of challenges while planting a garden and we can go from hope to hopelessness pretty quickly. Yet that is a normal part of life, so gardening is an easy place to start practicing these skills.

Say you find some tough ground you need to dig into to plant, you may get frustrated and give up. It is a good time to practice a **Stress Skill** like a 90-second pause or deep breathing, to calm down your stress response. Then try again! You may also notice when others get frustrated and teach them how to use this skill to navigate from their downstairs brain back upstairs.

Eating the sunflower seeds (if ok with your doctor) might be a good way for you to practice your **Happiness Habits.** Sunflower seeds are nutritious, high in choline and selenium, great for brain function and memory. You might also get some exercise planting gardens, and spend time in nature, two other Happiness Habits and great ways to release endorphins.

Planting gardens remind us to take **Inspired Actions** by setting specific goals for the garden. If we want a garden, we need to set a SMART goal about how many flowers, when and where we want the garden, and how we are going to grow the flowers. It is best if we write down the plan, chunk it down into actionable steps, think about obstacles and multiple ways we might overcome them, and check in with someone regularly to ensure progress.

We can cultivate our **Nourishing Networks** by planting gardens with others. That way, if we have challenges while planting, we can face them together and be more creative about overcoming them. And if we don't live by the person we want to plant with, we can both decide to plant and check in regularly on the garden. It is also super fun to plan community gardens, or even fields of sunflowers, and all join together in learning and practicing skills to Shine Hope.

And finally, time to get serious about **Eliminating Challenges**. For example, if our sunflowers die and we fail for a season of planting, it is easy for us to think of ourselves as failures. Yet we aren't failures, our process failed. So deconstruct the process. Did we under or over water? Did we plant at the wrong time of year? Was something wrong with the soil? Did we overwater? It is time to investigate, and instead of ruminating about the sunflowers start figuring out what we can do better to try again next year.

Planting sunflowers is a way to spread the message of hope, as if you put up a Gardens of Hope sign with the website, people can then find the curriculum to learn more about the programs for 'how' to hope. Our program is available around the world, and gardens are a great way to share the message that Hope is Teachable.

Find out more at www.hopefulcities.org

@theshinehopecompany     

Branding Guidelines

We've created guidelines for you so that you can customize this material to make it your own, yet also create consistency in messaging and symbolism. As consistent, repeated use of the symbol is the way we create the brand, and it is critical to do so if we want a global movement for Hope. That way, when used around the world, it is iconic, recognizable, and consistent, which helps to reinforce the Science of Hope and the skills necessary to Shine Hope.

We've included branding guidelines in the Appendix *(Appendix pg. 92)*, and we ask a few things of you:

- Use our trademarks consistently and with an approved license from us. Email us at activate@theshinehopcompany.com to obtain licenses and approvals. While some materials are available at no cost, there is a fee associated with becoming an official Hopeful City, depending on the size of the city. We do work with you to obtain sponsors if costs are an issue.
- Adhere to copyright laws.
- Contact us to get permission to change materials.
- Adhere to the yellow color scheme; we've provided PMS and CMYK colors for matching in the branding guidelines in the Appendix.
- Utilize our materials as much as possible. While not required, it adds to the consistency in messaging across cultures and communities, allows for imprinting of messaging, and adds to the Global Movement for Hope.
- Provide feedback so we can improve.

If you have any questions about them or concerns about their usage, reach out to us at activate@theshinehopecompany.com.

You can download the Branding Guidelines here or Scan the QR Code:
www.Hopefulcities.org/branding-guidelines/

Creating A Global Movement for Hope

Creating a Global Movement for Hope is among the most important components in ensuring everyone learns about Hope Science. To make this movement a reality and ensure that everyone has equitable access to Hope, many of the tools presented in this guide are no cost to use and share. **We have noted what is no cost, yet if you use them please do not alter without our permission.**

We've identified five keys to Shine Hope that help individuals and communities create, maintain, and grow Hope. They include:

(S) tress Skills

Identifying and proactively managing the stress response

(H) appiness Habits

Performing healthy long-term actions that foster positive feelings

(I) nspired Actions

Setting meaningful goals to nurture passion and purpose

(N) ourishing Networks

Building social connections that cultivate Hope

(E) liminating Challenges

Stopping negative thinking patterns that get in the way of Hope, such as worry, rumination, internalizing failure, limiting beliefs, negative bias, and trying to control things outside of our control.

To create a Global Movement for Hope and activate Hope in your community, it is important to engage the diverse groups in your area, including those in Science, Arts, Government, Education, Nonprofit, Housing, Police, Hospitals, and Business. In this guide, we share ways to activate these sectors.

To become an official Hopeful City and utilize trademarks, please reach out at activate@theshinehopecompany.com.

SECTION 2

Eight Guiding Principles

Eight Guiding Principles

Our guiding principles our embedded into every one of our interventions across all six sectors. We use the following guiding principles to drive change and ensure everyone can proactively move from Hopelessness to Hope.

Take a Whole Community Approach

Bridge the Knowledge-Action Gap

Utilize Solution-Focused Methods

Act Early Everywhere

Empower All Citizens

Amplify a Universal Brand

Use Evidence-Informed until Evidence-Based

Activate the Shine framework

Take a Whole Community Approach

Everyone experiences moments of Hopelessness. Every single person in a city. It is how we manage those moments that matter.

That is the difference between our strategy and a standard mental health strategy. We don't diagnose people or wait until they are in jail to get them support. We ensure everyone knows what Hopelessness is, and inoculate them with tools to Shine Hope. Everyone needs to learn Hope as a skill, so all of our programming is inclusive and impacts everyone.

We can not predict who will die by suicide, develop an addiction, or become violent other than if they are Hopeless, as it is the single consistent predictor and we all experience it. So what we do is take a whole community approach, and make Hope everyone's business by engaging community leaders, workplaces, educators, children, and all members of the community with a common language and framework for 'how' to Hope.

We all need Hope, and we all need to know how to Hope.

Some of the key initiatives we suggest to Take a Whole Community Approach include using the Shine framework everywhere, launching a citywide public health campaign, and sharing our educational programming.

Bridge the Knowledge-Action Gap

To accomplish a whole city approach, leaders and community members need to simultaneously share their own moments of Hopelessness while sharing how they use The Five Keys to Shine Hope.

Some of the key initiatives we suggest to Bridge the Knowledge-Action Gap include, encouraging government officials, CEOS, and presidents to create their own 'Hope Story" by using the Shine framework and the 20/80 rule, conducting research on Hope, and using a common brand for Hope.

Utilize Solution-Focused Methods

So much work these days is focused on the problems rather than the scope of the issue. We encourage all, throughout all of our programs, to focus on solutions. In our Hope Hero Stories, and My Hope Stories, we suggest all spend 20% of the time on the problem, 80% on the solution.

So much work these days is focused on the problems rather than the scope of the issue. We encourage all, throughout all of our programs, to focus on solutions. In our Hope Hero Stories, and My Hope Stories, we suggest all spend 20% of the time on the problem, 80% on the solution.

It is time we look to others that have navigated similar problems, and see how they utilized the Shine framework to overcome them. We study individuals, and encourage others to do so. We then encourage all to put those skills into practice.

The more we focus on the problem, the bigger it gets. We aim to change that conversation to the solution.

Some of the key initiatives we suggest to Utilize Solution-Focused Methods include empowering all individuals that they can learn the skills to activate Hope, and sharing the message of Hope through My Hope Hero and My Shine Hope Story™.

Act Early Everywhere

Equipping everyone with the skills on 'how' to Hope to ensure they can proactively move from Hopelessness to Hope before they experience the adverse outcomes of persistent Hopelessness. It is important to ensure all know what tools are available for a crisis before they experience a crisis, and are equipped to better manage change and adversity.

Some of the key initiatives to Act Early Everywhere include incorporating Hope education materials into schools for all children and youth, equipping parents with Hope skills, and infusing the workplace with messaging around Hope.

Empower All Citizens

Everyone has the right to learn the skills to Hope. Furthermore, activating Hope creates self-efficacy for your citizens, so they can start solving problems in the city. The reality is, we need an all hands approach to tackling our toughest issues, and it is impossible to do unless the citizens have Hope.

Some key initiatives to Empower All Citizens include donating resources to ensure all have access to the skills to Hope through programming, and working to reach all citizens within a city.

Amplify a Universal Brand

The impact of branding is crucial for establishing an easy to identify common language and framework. We now have attention spans of less than a goldfish, and it is hard to retain information. Consistent, recognizable repeated messaging is crucial for people to retain the information, which is why we believe transforming humanity from Hopelessness to Hope must be done through the power of a brand.

Some key initiatives to Amplify a Universal Brand include using consistent visual branding and standardizing communication around Hope.

Use Evidence-Informed until Evidence-Based

All of our programs began with evidence-informed approaches before they became evidence-based. We partner with Universities and researchers that want to study programs. Everything we do uses evidence, and we study as we go to improve and iterate, using science-backed approaches and building and publishing evidence as we go.

Some key initiatives to Use Evidence-Informed and Evidence-Based information about Hope is through ongoing data collection efforts, partnership to conduct studies, and encouraging everyone to measure their Hope using the Snyder Hope Scales.

Activate the Shine Framework

Empowering individuals with self-management skills and fostering peer support is crucial. Education is key–people can't address what they don't know. It is normal to struggle with emotions like sadness, anger, or fear; however, preemptive skill-building is essential to prevent the development of chronic adverse emotionality and subsequent outcomes. We empower people to channel these emotions into collective change.

Some key initiatives to Activate the Shine Framework include sharing digital brochures, infographics, and posters everywhere and spreading the message through comprehensive and strategic public health campaigns.

The Hopeful Cities® Playbook is a resource designed to make Hope accessible and operational in any city around the world. It is a step-by-step guide to help cities activate the "how-to" of Hope everywhere: workplaces, communities, schools, homes, and more. It is an actionable marketing plan that operationalizes Hope as it creates awareness about the Science of Hope.

You cannot wish for a Hopeful City. Hope takes action. We've outlined ways to activate Hope in your community by building a global brand and movement for Hope. It isn't enough to share a message once. We must repeatedly, consistently, and intentionally repeat messages so people know that no matter what life brings, there is always a way to move from Hopelessness to Hope.

We believe that Hope is a universal right. Therefore, we provide many of these resources at no cost and have left blank spaces on the materials so that you can include local sponsors. Yet if you want to license materials and become an official Hopeful City, there are costs. The Playbook teaches you how to partner with local businesses to engage and activate them in your community, and we have done our best to make these materials comprehensive, low-cost, and many at no cost to make Hope accessible to all.

If you want to sponsor a city activation, reach out at activate@theshinehopecompany.com.

SECTION 3

Six Sectors for Fostering Hope

OVERVIEW

SECTOR | **INTERVENTIONS**

GOVERNMENT

No Cost

1. International Day of Hope
2. Five-Day Hope Challenge & Social Media Campaign
3. My Shine Hope Story™ Template for Mayors
4. Digital Shine Hope Communication Assets: Shine Hope Infographic, Moments of Hope Cards, Teen and Adult Brochures, Posters, Signs, and Billboards

Fee for Programs

5. City Landing Page Listing Resources
6. Publicity Plan with Assets (Press Release, Staff Communications, Press Talking Points, Articles, and Public Service Announcement Script)
7. Print Shine Hope Communication Assets: Moments of Hope Cards, Teen and Adult Brochures, Posters, Signs, and Billboards
8. Hopeful Mindsets Workplace Overview Video Course License for use with City Staff
9. Management & Leadership Training

SCIENCE

No Cost

1. International Day of Hope
2. Five-Day Hope Challenge & Social Media Campaign
3. Digital Shine Hope Communication Assets: Shine Hope Infographic, Moments of Hope Cards, Teen and Adult Brochures, Posters, Signs, and Billboards
4. Measure Hope Scores
5. Feedback Mechanisms to Improve Hope Science
6. Online repository of Hope Science Research

Fee for Programs

7. Partnerships for Data Collection and Analysis
8. Print Shine Hope Communication Assets: Moments of Hope Cards, Teen and Adult Brochures, Posters, Signs, and Billboards

OVERVIEW

SECTOR	INTERVENTIONS

EDUCATION

No Cost

1. International Day of Hope
2. Five-Day Hope Challenge & Social Media Campaign
3. My Shine Hope Story™ Template & My Hope Hero Template
4. Digital Shine Hope Communication Assets: Shine Hope Infographic, Moments of Hope Cards, Teen and Adult Brochures, Posters, Signs, and Billboards
5. Hopeful Minds Overview K-6 Digital Version
6. Hopeful Minds Parent's Guide Digital Version

Fee for Programs

7. Hopeful Minds Overview K-6 Print Version
8. Hopeful Minds Parent's Guide Print Version
9. Hopeful Minds Deep Dive K-6 Digital Version and Print Version
10. Hopeful Minds Teen Hopeguide Digital Version and Print Version
11. Hopeful Mindsets Workplace Overview Video Course License for use with Teachers
12. Hopeful Mindsets on the College Campus 10-Module Video Course
13. Print Shine Hope Communication Assets: Moments of Hope Cards, Teen and Adult Brochures, Posters, Signs, and Billboards
14. Hope Ambassador Program & Hope Clubs Framework
15. Community Members Teaching Hope Video Training (police, older population, etc.)
16. Educator Training

WORKPLACE

No Cost

1. International Day of Hope
2. Five-Day Hope Challenge & Social Media Campaign
3. Digital Shine Hope Communication Assets: Shine Hope Infographic, Moments of Hope Cards, Teen and Adult Brochures, Posters, Signs, and Billboards
4. Workplace Research
5. Cause Marketing Campaign
6. Employee Volunteer Opportunities
7. Hopeful Minds Parent's Guide Digital Version
8. Hopeful Minds Overview K-6 Digital Version
9. My Shine Hope Story™ for Leadership

OVERVIEW

SECTOR | **INTERVENTIONS**

WORKPLACE

Fee for Programs

10. Workplace Leadership & Management Training
11. Hopeful Mindsets Workplace Overview Video Course License
12. Workplace Keynote Presentation at Conferences
13. Print Shine Hope Communication Assets: Moments of Hope Cards, Teen and Adult Brochures, Posters, Signs, and Billboards
14. Hopeful Minds Parent's Guide Print Version
15. Hopeful Minds Overview K-6 Print Version

HEALTHCARE

No Cost

1. International Day of Hope
2. Five-Day Hope Challenge & Social Media Campaign
3. Digital Shine Hope Communication Assets: Shine Hope Infographic, Moments of Hope Cards, Teen and Adult Brochures, Posters, Signs, and Billboards
4. Hopeful Minds Parent's Guide Digital Version
5. Hopeful Minds Overview K-6 Digital Version
6. My Shine Hope Story™ for Leadership

Fee for Programs

7. Workplace Leadership & Management Training
8. Workplace Keynotes
9. Hopeful Mindsets Overview Workplace Video Course License
10. Hopeful Mindsets General Overview Video Course License for Patients
11. Printed Shine Hope Educational Materials for Staff and Waiting Rooms: Moments of Hope Cards, Teen and Adult Brochures, Posters, Signs, and Billboards
12. Workplace & Patient Research
13. Online Volunteer Training
14. Hopeful Minds Parent's Guide Print Version
15. Hopeful Minds Overview K-6 Print Version

OVERVIEW

SECTOR | **INTERVENTIONS**

ART

Note: All art features a template that showcases the Shine Hope framework and a link to where to go in the community for support to drive awareness and serve as prevention of crisis. We have templates for you to put up mural signs, murals, gardens, artwork, etc. No materials or supplies are provided.

No Cost

1. International Day of Hope
2. Shine Hope Sunflower Mural Templates
3. Sunflower Fascinating Facts Digital Poster
4. Sunflower Gardens for Hope
5. Sunflower Fields for Hope
6. Public Art Hope Health Campaigns

Fee for Programs

7. Hope Heroes Exhibit
8. Interactive Art Educational Experiences
9. Iconic Large Hope Letters

Hope is like a muscle—we're doing all we can to activate hope around the city and grow it together.

- Mayor Hillary Schieve, Reno, NV

Government

Incorporating government officials into Hope activation initiatives is pivotal for fostering widespread change and resilience within communities.

Their involvement sets powerful examples by normalizing discussions on Hope and mental health, influencing policy decisions that allocate resources toward Hope-building programs, and amplifying awareness through public campaigns. Government leadership fosters collaboration among diverse community groups and ensures the integration of Hope-focused strategies into long-term policies, guaranteeing sustained efforts toward empowering individuals, families, and entire communities to navigate challenges with optimism and resilience.

Government officials must recognize Hope as a strategic imperative due to its profound impact on society's well-being. Embracing Hope as a central strategy can lead to reduced mental health issues, fostering a more resilient and cohesive community fabric. By investing in Hope-building initiatives, officials can enhance workforce productivity, reduce societal challenges (i.e., violence and crime), and potentially decrease long-term costs associated with mental health care.

Moreover, integrating Hope into policies ensures more effective, sustainable solutions to societal issues, creating a pathway toward a brighter future for generations to come. Prioritizing Hope sends a powerful message of commitment to citizen well-being, fostering trust and active engagement within communities. It's not just about addressing immediate concerns but about shaping a more Hopeful and resilient society for the long term.

TAKE ACTION WITH THE FOLLOWING INTERVENTIONS

(No cost)

1. International Day of Hope

The International Day of Hope, scheduled on July 12th, provides an opportunity for global solidarity. On this day, we come together to share the science, stories, and strategies of Hope and actively engage in implementing Hope in our lives and communities worldwide. Our goal is to advocate for the establishment of The International Day of Hope through an official United Nations resolution.

This initiative will kick off a five-day campaign featuring the Five-Day Global Hope Challenge, yard signs, sunflower gardens, workplace educational posters, Shine Hope sunflower murals, live speaking events, classroom teachings of Hopeful Minds, and more.

We encourage Mayors, Governors, and schools to issue proclamations for their towns and environments, demonstrating solidarity in officially recognizing the day and joining the Hopeful Cities Movement.

Day 1
The Impact of Hopelessness, and the Science of Hope
#GlobalHopeChallenge
Be Part Of Activating Hope Around The World.

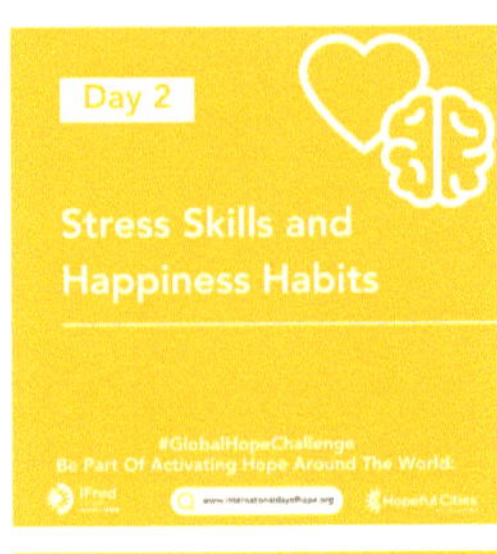
Day 2
Stress Skills and Happiness Habits
#GlobalHopeChallenge
Be Part Of Activating Hope Around The World.

Day 3
Inspired Actions
#GlobalHopeChallenge
Be Part Of Activating Hope Around The World.

2. Five-Day Hope Challenge & Social Media Campaign

The Five-Day Global Hope Challenge is a five-day e-mail challenge ensuring all know the what, why, and how to Hope. It reviews what Hopelessness is, the Shine Hope framework, and instructs how to measure Hope. It is a simple way to get started learning how to Shine Hope.

Social media is a great way to share the resources available for Hope with your friends, family, and community, so we have created a social media toolkit for Hope as well. All of our images and content are available to download at no cost to share and activate the message for Hope. Download the Hope Challenge Social Media Kit for daily social media posts: Hopefulcities.org/social-media/

You can also help your community access the Hopeful Cities resources by tagging us in your posts using @theshineHopecompany @ifredorg #HopefulCities #Hope

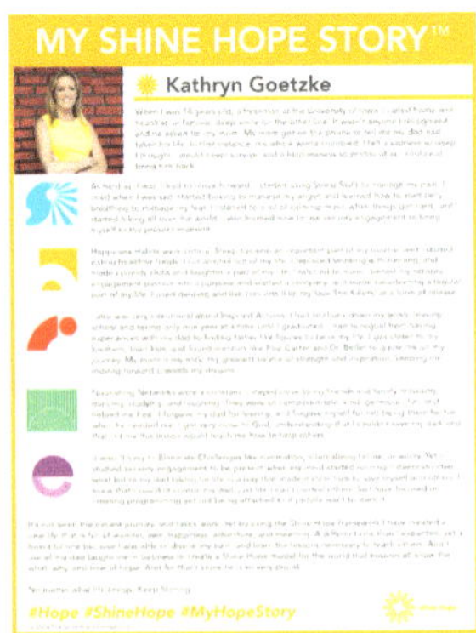
MY SHINE HOPE STORY™
Kathryn Goetzke
#Hope #ShineHope #MyHopeStory

3. My Shine Hope Story™ Template for Mayors

My Shine Hope Story™ gives community leaders an avenue to normalize Hopelessness and share their unique experiences by using the Shine Hope framework. We all experience moments of Hopelessness (i.e., emotional despair and motivational helplessness). How we manage the moments of Hopelessness matters.

We suggest you share a recent challenge, large or small. We encourage you to spend 20% of the time describing what the Hopelessness was about (sadness, anger, fear, and powerlessness). We then ask you to spend 80% of the story sharing how you overcame it.

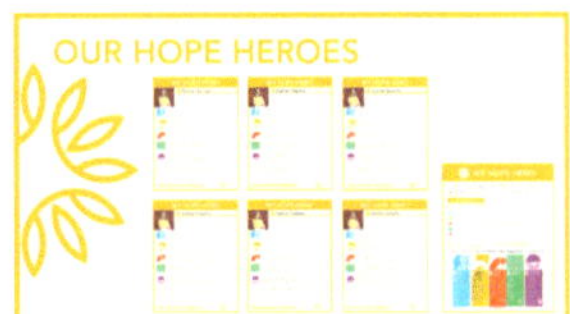

What Stress Skills did you use? What Happiness Habits did you practice? What Inspired Actions did you take to make it through? How did your Nourishing Networks support you? And what was one of your biggest challenges (i.e., negative thought pattern), and how did you overcome it?

We created an outline for you to use while creating your Hope Story, and you can use our Shine Hope infographic for some ideas. You can download the Shine Hope Infographic here: www.theshinehopecompany.com/shine-Hope/

We encourage you to share your story with the community while encouraging others to share their stories as well. Discuss what skills work, help all practice, and be sure people know where to go for support if they are having challenges in your city. You can download the My Shine Hope Story™ Template here: www.Hopefulcities.org/government.

If you want to share your Hope story on social media, we encourage you to do so. Please tag us @ifredorg @theshineHopecompany #myHopestory #myHopehero so we can share with others.

4. Digital Communication Assets: Shine Hope Infographic, Moments of Hope Cards, Teen and Adult Brochures, Posters, Signs, and Billboards

Our digital communications provide an opportunity to educate all on the Shine Hope framework in an easy-to-understand way. We have an infographic and digital downloadable tool for newsletters or social media posts to describe the Shine Hope framework at no cost.

We have two brochures that are great for sharing in newsletters or other city-wide communications. Our Digital Shine Hope Brochure is geared toward adults, while our Teen Shine Hope brochure is similar, yet more teen-friendly. Both brochures include tips on the Shine Hope framework, and have resources to help elevate Hope in your life and with others.

We also have Shine Hope Posters you can use as individual images, or on web portals, that give more details into the Shine Hope Framework. Lastly, we have a Sunflower Fascinating Facts poster for some interesting tidbits on sunflowers, if you have an interest!

Get your digital copies here: theshinehopecompany.com/shine-Hope/

See the appendix for billboards, posters, and signs that you can print or purchase to incorporate into your educational campaign about 'how' to Hope.

(Programs with Fees)

5. City Landing Page Listing Resources

As an official Hopeful City partner, we create a landing webpage with your city's resources. The landing page will include a list of community resources that you provide to us, or we can hyperlink to a page on your city's website that already contains resources. The goal of the landing page is to ensure that everyone in the community has one place with a list of resources they can use if they are having challenges with things like housing, mental health, finances, crises, and more.

It is listed as: www.Hopefulcities.org/country/state/city

Your city gets added to our website, and as people see the web link to www.Hopefulcities.org they see your webpage. We are also able to add a specific QR code for your webpage if you want, which you can also add to communications throughout your city, such as press releases.

The SHINE Hope Company Activates Workplace Mental Health at SHRM

6. Publicity Plan with Assets (Press Release, Staff Communications, Press Talking Points, Articles, and Public Service Announcement Script)

A publicity plan is key to the success of implementing the Hopeful City Playbook in your community, as it spreads the message of Hope. We've tried to make it easy for you, and included a press release template highlighting the launch of your city as part of the Hopeful Cities movement, and why it is important. We've also included articles for communications, Public Service Announcement (PSA) scripts for your leadership to use on TV or radio, and press talking points. The idea of the campaign is to ensure all know what Hopelessness is, that they can learn how to Hope, and become equipped with skills through the Shine Hope framework.

7. Print Shine Hope Educational Moments of Hope Cards, Teen and Adult Brochures, Posters, Signs, and Billboards

Our print communications provide an opportunity to educate all on the Shine Hope framework in an easy-to-understand way. We want to ensure all are equipped with a crisis hotline and skills to proactively activate their Hope. It is our goal to ensure all know that Hope is a skill, measurable, and teachable.

We have two brochures, a general one and a teen-focused one, available for waiting rooms, community centers, lobbies, offices, or libraries. We have posters for the walls in workplaces, coffee shops, libraries, or waiting rooms, and Moments of Hope cards to hand out anywhere people gather. Lastly, we have a Sunflower Fascinating Facts poster for some interesting tidbits on sunflowers, if you have an interest! All of these print materials serve as a cost-effective approach to getting the word out about Hope, and start equipping the population to take a proactive approach to managing Hope in their life and the lives of others.

You can purchase printed versions of the Shine Hope Educational materials here: theshineHopestore.com/collections/spread-Hope

See the appendix for billboards, posters, and signs that you can print or purchase to incorporate into your educational campaign about 'how' to Hope.

8. Hopeful Mindsets Workplace Overview Video Course License for use with City Staff

The Hopeful Mindsets Workplace Overview is a 90-minute video course that introduces Hope and the Five Keys to Shine Hope™ framework to help you create, maintain, and grow Hope in the workplace. We give an overview of the framework, so you can then apply it to your career to activate Hope at work. The course is available to license to entire companies, to ensure all know the what, why, and how of Hope.

The video course is available for individual purchase at www.Hopecourses.com. For bulk license, e-mail us at activate@theshinehopecompany.com.

9. Management & Leadership Training

We offer diverse options for management and leadership training, incorporating the measurement of managers' Hope levels and the use of the VIA Strength Finder—a powerful tool that identifies strengths to optimize overall leadership performance.

This training covers various aspects, including insights into the Hope Matrix, an exploration of the definition and consequences of Hopelessness, and the provision of strategies for proactively managing moments of Hopelessness while instilling the necessary skills to activate Hope. Managers are specifically guided on the effective application of the Shine Hope framework in their leadership roles, integrating our Hopeful Minds Overview for the Workplace workbook. They are encouraged to explore how Hope can serve as inspiration and support for their teams.

The primary goal is to equip company leaders with essential skills and ensure their awareness of available resources before reaching a crisis point. The program is delivered through a 90-minute in-person meeting, fostering a collaborative environment that emphasizes learning and practical application.

For executive leadership, high-level Leadership Training is offered to provide a profound understanding of the science of Hope. This training introduces the Shine Hope framework, offering strategies and techniques to nurture Hope in the workplace.

It includes a presentation tailored to industry-specific statistics. Your leadership team will gain insights into the psychology of Hope, measurement methods, industry-specific Hope considerations, cultivating Hopeful mindsets, and the practical application of Hope in leadership. The executive leadership training session is designed to meet the needs of top leaders and their busy schedules and thus are shorter (i.e., 15- 60 minutes); however, the training can be tailored according to needs.

To complement management and leadership training, the Hopeful Mindsets Workplace Overview video course serves as an excellent addition, ensuring that entire companies understand the what, why, and how of Hope.

To learn more about management and leadership training, please email us at **activate@theshinehopecompany.com**

Science

Science is crucial for activating Hope within a city by providing empirical evidence, research-backed strategies, and data-driven insights that underpin effective interventions.

Through scientific inquiry, cities can develop a nuanced understanding of Hope, its impact on mental health, and the efficacy of diverse interventions. Scientifically informed approaches enable the identification of effective practices tailored to diverse populations, ensuring inclusivity and accessibility. Moreover, social psychologists have found that people are more likely to have positive attitudes about something if the topic is well-cited; thus, using science alongside the Hope campaign increases the likelihood of engagement.

Furthermore, by conducting research, you are adding to Hope Science and to our overall understanding of Hope, which equips individuals and communities with the tools needed to navigate challenges and setbacks, ultimately paving the way for a more Hopeful and flourishing cityscape.

TAKE ACTION WITH THE FOLLOWING INTERVENTIONS

(No cost)

1. International Day of Hope

The International Day of Hope, scheduled on July 12th, provides an opportunity for global solidarity. On this day, we come together to share the science, stories, and strategies of Hope and actively engage in implementing Hope in our lives and communities worldwide. Our goal is to advocate for the establishment of The International Day of Hope through an official United Nations resolution.

This initiative will kick off a five-day campaign featuring the Five-Day Global Hope Challenge, yard signs, sunflower gardens, workplace educational posters, Shine Hope sunflower murals, live speaking events, classroom teachings of Hopeful Minds, and more.

We encourage Mayors, Governors, and schools to issue proclamations for their towns and environments, demonstrating solidarity in officially recognizing the day and joining the Hopeful Cities Movement.

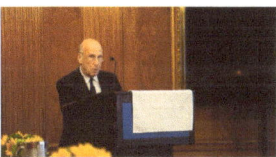

2. Five-Day Hope Challenge & Social Media Campaign

Day 1
The Impact of Hopelessness, and the Science of Hope
#GlobalHopeChallenge
Be Part Of Activating Hope Around The World

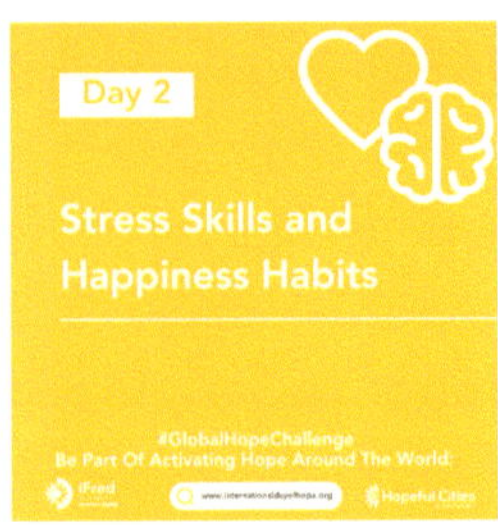
Day 2
Stress Skills and Happiness Habits
#GlobalHopeChallenge
Be Part Of Activating Hope Around The World

Day 3
Inspired Actions
#GlobalHopeChallenge
Be Part Of Activating Hope Around The World

The Five-Day Global Hope Challenge is a five-day e-mail challenge ensuring all know the what, why, and how to Hope. It reviews what Hopelessness is, the Shine Hope framework, and instructs how to measure Hope. It is a simple way to get started learning how to Shine Hope.

Social media is a great way to share the resources available for Hope with your friends, family, and community, so we have created a social media toolkit for Hope as well. All of our images and content are available to download at no cost to share and activate the message for Hope. Download the Hope Challenge Social Media Kit for daily social media posts: Hopefulcities.org/social-media/

You can also help your community access the Hopeful Cities resources by tagging us in your posts using @theshineHopecompany @ifredorg #HopefulCities #Hope

3. Digital Communication Assets: Shine Hope Infographic, Moments of Hope Cards, Teen and Adult Brochures, Posters, Signs, and Billboards

Do you know how to
#SHINEHOPE

These are signs for
HOPE

HOPE is Teachable
HOPE is Teachable

Our digital communications provide an opportunity to educate all on the Shine Hope framework in an easy-to-understand way. We have an infographic and digital downloadable tool for newsletters or social media posts to describe the Shine Hope framework at no cost.
We have two brochures that are great for sharing in newsletters or other city-wide communications. Our Digital Shine Hope Brochure is geared toward adults, while our Teen Shine Hope brochure is similar, yet more teen-friendly. Both brochures include tips on the Shine Hope framework, and have resources to help elevate Hope in your life and with others.

We also have Shine Hope Posters you can use as individual images, or on web portals, that give more details into the Shine Hope Framework. Lastly, we have a Sunflower Fascinating Facts poster for some interesting tidbits on sunflowers, if you have an interest!

Get your digital copies here:
theshinehopecompany.com/shine-Hope/

See the appendix for billboards, posters, and signs that you can print or purchase to incorporate into your educational campaign about 'how' to Hope.

4. Measure Hope Scores

We use Children and Adult Snyder Hope Scales, which are robust scientific measures that are used to study the outcomes of Hope we share in this program. Hope is measurable and higher Hope can lead to better outcomes. Encourage all those in your community to measure Hope, so we can start tracking Hopefulness in individuals around the world.

You can find the free scales, and encourage others to take them, here: theshinehopecompany.com/measure-your-Hope/

5. Feedback Mechanisms to Improve Hope Science

Feedback helps us learn more about the science of Hope and it helps us adapt our programming to ensure it's equitable. The more you do to get involved in Hope Science and contribute to our work, the more it will be improved. We ask that you send us feedback using this feedback form. We also include feedback surveys within each of our programs and ask that those who complete our programs complete those surveys to help us understand how to further improve. The more we know, the better we do.

6. Online Repository of Hope Science Research

We are compiling research on Hope around the world to add to our collective knowledge, which gives a snapshot view of all the positive outcomes linked to Hope. You can check out our research page, and see the latest in Hope science. If you know of a Hope study, send it over to veronica@theshinehopecompany.com so we can add it to the database.

You can access the research page here: theshinehopecompany.com/Hope-science/general/

We talk a lot about agency when we talk about Hope using Snyder's approach, and so agency is really self efficacy [...] this power of self belief.

- Dr. Crystal Bryce, PhD

(Programs with Fees)

7. Partnerships for Data Collection and Analysis

We've partnered with research institutions around the world to study the effectiveness of our programming, and encourage you to ask your research institutions and universities to join us. We are always looking for research partners to conduct studies on the effectiveness of Hope interventions within diverse populations. We aim to continue to improve and tailor strategies to address specific community needs.

Consider working with us to create custom programs, research specific populations, or create additional studies on current programs. You can review the research we have completed on the Hopeful Minds curriculums at theshinehopecompany.com/research/

If you wish to partner with us, please contact us at activate@theshinehopecompany.com

8. Print Shine Hope Educational Moments of Hope Cards, Teen and Adult Brochures, Posters, Signs, and Billboards

Our print communications provide an opportunity to educate all on the Shine Hope framework in an easy-to-understand way. We want to ensure all are equipped with a crisis hotline and skills to proactively activate their Hope. It is our goal to ensure all know that Hope is a skill, measurable, and teachable.

We have two brochures, a general one and a teen-focused one, available for waiting rooms, community centers, lobbies, offices, or libraries. We have posters for the walls in workplaces, coffee shops, libraries, or waiting rooms, and Moments of Hope cards to hand out anywhere people gather. Lastly, we have a Sunflower Fascinating Facts poster for some interesting tidbits on sunflowers, if you have an interest! All of these print materials serve as a cost-effective approach to getting the word out about Hope, and start equipping the population to take a proactive approach to managing Hope in their life and the lives of others.

You can purchase printed versions of the Shine Hope Educational materials here: theshineHopestore.com/collections/spread-Hope

See the appendix for billboards, posters, and signs that you can print or purchase to incorporate into your educational campaign about 'how' to Hope.

Education

We often tell people to have Hope, yet Hope is not something that has typically been "taught." The word itself is often misused in the popular media, used more like a "wish" than something we control with action. So how can we expect people to move from Hopelessness to Hope, unless we teach them how?

This is why we are so passionate about teaching Hope, especially to kids. We know if we reach them at an early age (before 10), the skills we teach them are much easier to integrate. And by teaching Hope skills at an early age, we are equipping them with the roadmap they need to start reaching their goals.

Our programs meet both National Health Educational Standards (NHES) and Social and Emotional Learning (SEL) Guidelines. You can find all the information about how on the Hopeful Minds website and in the curriculums themselves. The programs are used across cultures, yet we suggest you adapt stories and Hope Heroes to make them more culturally specific.

It has an impact not just on the students, but the educators as well. Our program has been downloaded by more than 1500 educators around the world, with more accessing it every day. We have translated the student workbooks to Spanish, and are doing our best to get the programs translated to other languages as well. We welcome your support in making that happen.

As Hope is predictive of not just life outcomes, but school outcomes, we believe every child must be taught the "how to" of Hope. Their level of Hope impacts their ability to graduate, their engagement in class, their learning, and overall health, which all relate to a city's ability to function. Hope even predicts their sports performance, so we encourage you to inspire kids to Hope using athletes as examples.

"Every child in the world needs this program."

- Hopeful Minds Student from Northern Ireland

TAKE ACTION WITH THE FOLLOWING INTERVENTIONS

(No cost)

1. International Day of Hope

The International Day of Hope, scheduled on July 12th, provides an opportunity for global solidarity. On this day, we come together to share the science, stories, and strategies of Hope and actively engage in implementing Hope in our lives and communities worldwide. Our goal is to advocate for the establishment of The International Day of Hope through an official United Nations resolution.

This initiative will kick off a five-day campaign featuring the Five-Day Global Hope Challenge, yard signs, sunflower gardens, workplace educational posters, Shine Hope sunflower murals, live speaking events, classroom teachings of Hopeful Minds, and more.

We encourage Mayors, Governors, and schools to issue proclamations for their towns and environments, demonstrating solidarity in officially recognizing the day and joining the Hopeful Cities Movement.

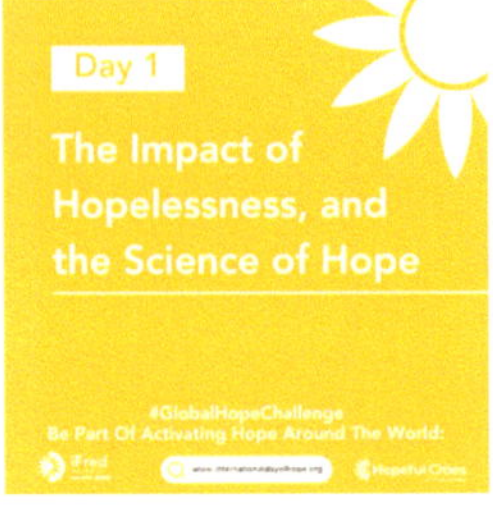
Day 1
The Impact of Hopelessness, and the Science of Hope
#GlobalHopeChallenge
Be Part Of Activating Hope Around The World:

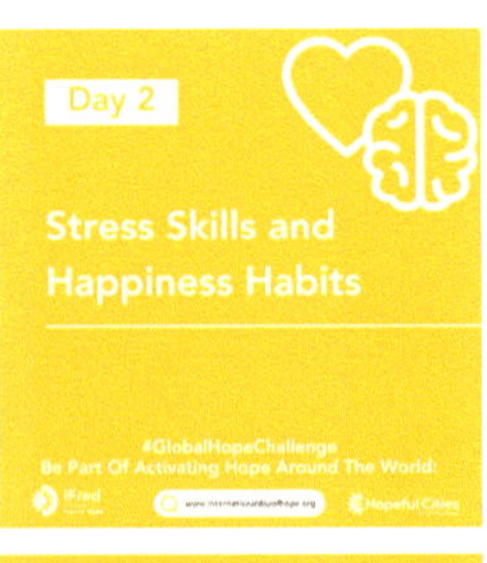
Day 2
Stress Skills and Happiness Habits
#GlobalHopeChallenge
Be Part Of Activating Hope Around The World:

Day 3
Inspired Actions
#GlobalHopeChallenge
Be Part Of Activating Hope Around The World:

2. Five-Day Hope Challenge & Social Media Campaign

The Five-Day Global Hope Challenge is a five-day e-mail challenge ensuring all know the what, why, and how to Hope. It reviews what Hopelessness is, the Shine Hope framework, and instructs how to measure Hope. It is a simple way to get started learning how to Shine Hope.

Social media is a great way to share the resources available for Hope with your friends, family, and community, so we have created a social media toolkit for Hope as well. All of our images and content are available to download at no cost to share and activate the message for Hope. Download the Hope Challenge Social Media Kit for daily social media posts: Hopefulcities.org/social-media/

You can also help your community access the Hopeful Cities resources by tagging us in your posts using @theshineHopecompany @ifredorg #HopefulCities #Hope

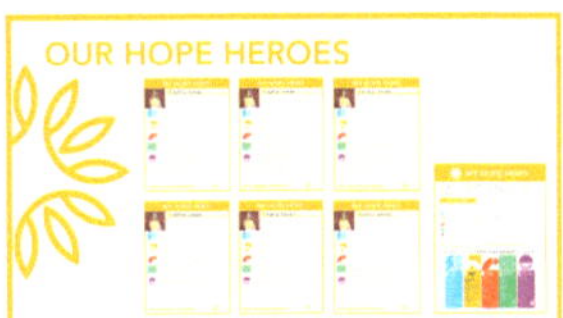

3. My Shine Hope Story™ & My Hope Hero Template

My Shine Hope Story™ gives community leaders an avenue to normalize Hopelessness and share their unique experiences by using the Shine Hope framework. We all experience moments of Hopelessness (i.e., emotional despair and motivational helplessness). How we manage the moments of Hopelessness matters.

We suggest you share a recent challenge, large or small. We encourage you to spend 20% of the time describing what the Hopelessness was about (sadness, anger, fear, and powerlessness). We then ask you to spend 80% of the story sharing how you overcame it.

What Stress Skills did you use? What Happiness Habits did you practice? What Inspired Actions did you take to make it through? How did your Nourishing Networks support you? And what was one of your biggest challenges (i.e., negative thought pattern), and how did you overcome it?

We created an outline for you to use while creating your Hope Story, and you can use our Shine Hope infographic for some ideas. You can download the Shine Hope Infographic here: theshinehopecompany.com/shine-Hope/

We encourage you to share your story with the community while encouraging others to share their stories as well. Discuss what skills work, help all practice, and be sure people know where to go for support if they are having challenges in your city. You can download the My Shine Hope Story™ Template here: www.Hopefulcities.org/education.

If you want to share your Hope story on social media, we encourage you to do so. Please tag us @ifredorg @theshineHopecompany #myHopestory #myHopehero so we can share with others.

4. Digital Communication Assets: Shine Hope Infographic, Moments of Hope Cards, Teen and Adult Brochures, Posters, Signs, and Billboards

Our digital communications provide an opportunity to educate all on the Shine Hope framework in an easy-to-understand way. We have an infographic and digital downloadable tool for newsletters or social media posts to describe the Shine Hope framework at no cost.

We have two brochures that are great for sharing in newsletters or other city-wide communications. Our Digital Shine Hope Brochure is geared toward adults, while our Teen Shine Hope brochure is similar, yet more teen-friendly. Both brochures include tips on the Shine Hope framework, and have resources to help elevate Hope in your life and with others.

We also have Shine Hope Posters you can use as individual images, or on web portals, that give more details into the Shine Hope Framework. Lastly, we have a Sunflower Fascinating Facts poster for some interesting tidbits on sunflowers, if you have an interest!

Get your digital copies here:
theshinehopecompany.com/shine-Hope

See the appendix for billboards, posters, and signs that you can print or purchase to incorporate into your educational campaign about 'how' to Hope.

5. Hopeful Minds Overview K-6 Digital Version

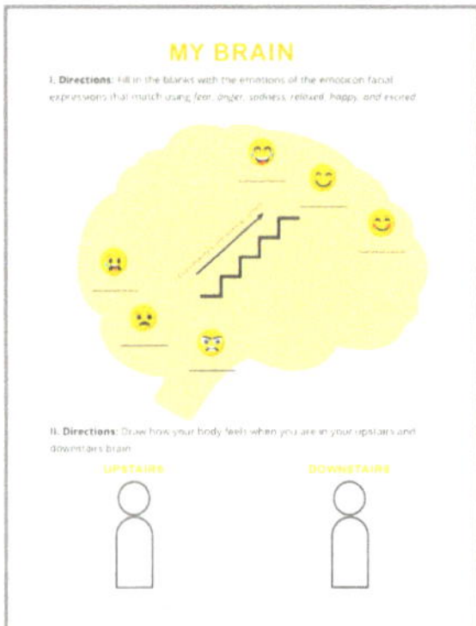

The Hopeful Minds Overview Educator Guide is a curriculum designed to give children an introduction to the "what," "why," and "how" of Hope. The curriculum includes three, one-hour lessons that introduce the key tools needed to create, maintain, and grow Hope. Additionally, the curriculum includes background information for educators, supplemental resources, classroom visuals, and a Hopework Book for students.

While this curriculum is geared towards 2nd-grade students, it can be easily adapted for any age range (adults included) and can be utilized in any setting (such as schools, after-school programs, places of worship, hospitals, offices, and more). This curriculum has been specially designed to be used for either classroom or remote learning.

The digital guide is available at no cost. This is the downloadable version of the curriculum, and we have downloadable fillable workbooks as well.

Get your copy here: www.Hopefulminds.org/curriculums

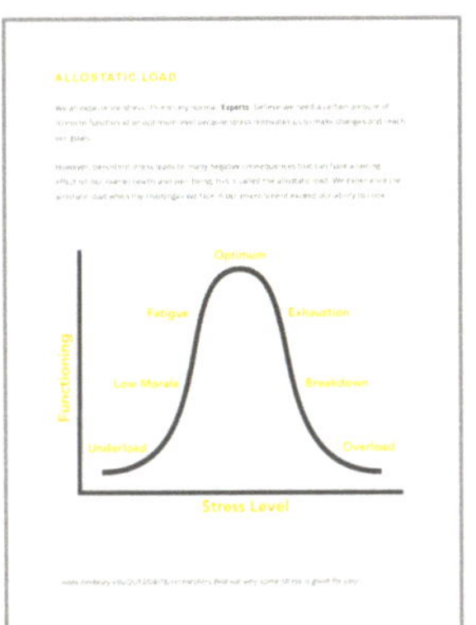

6. Hopeful Minds Parent's Guide Digital Version

The Hopeful Minds Parent's Guide provides a broad overview of the concepts discussed in the Hopeful Minds curriculums, and provides parents with easy ways to implement the Five Keys to Shine Hope (Stress Skills, Happy Habits, Inspired Actions, Nourishing Networks, and Eliminating Challenges) and Hopeful language in the home. It is helpful when parents reinforce Hope language at home, so the family can practice together. Hope is measurable and teachable. With Hopelessness at an all-time high in youth, we've got to be proactive about practicing these skills with youth. This is a tool parents can use to talk about mental health in a positive, proactive way at home and know what to look out for and find resources for support.

The digital guide is available at no cost.
Get your copy here: www.Hopefulminds.org/curriculums

(Programs with Fees)

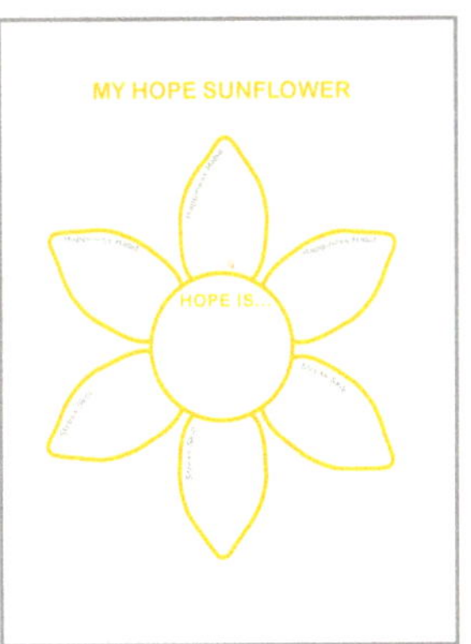

7. Hopeful Minds Overview K-6 Print Version

The Hopeful Minds Overview Educator Guide is a curriculum designed to give children an introduction to the "what," "why," and "how" of Hope. The curriculum includes three, one-hour lessons that introduce the key tools needed to create, maintain, and grow Hope. Additionally, the curriculum includes background information for educators, supplemental resources, classroom visuals, and a Hopework Book for students.

While this curriculum is geared towards 2nd-grade students, it can be easily adapted for any age range (adults included) and can be utilized in any setting (such as schools, after-school programs, places of worship, hospitals, offices, and more). This curriculum has been specially designed to be used for either classroom or remote learning.

The digital guide is available at no cost. This is the downloadable version of the curriculum, and we have downloadable fillable workbooks as well.

Get your copy here: www.Hopefulminds.org/curriculums

Some may prefer hard copies of our programming. Visit www.Hopefulminds.org/curriculums to find out where to order hard copies for personal use or order for your library or bookstore.

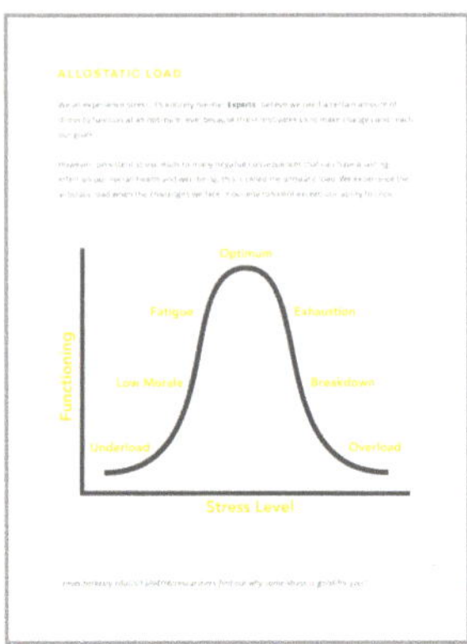

8. Hopeful Minds Parent's Guide Print Version

The Hopeful Minds Parent's Guide provides a broad overview of the concepts discussed in the Hopeful Minds curriculums, and provides parents with easy ways to implement the Five Keys to Shine Hope (Stress Skills, Happy Habits, Inspired Actions, Nourishing Networks, and Eliminating Challenges) and Hopeful language in the home. It is helpful when parents reinforce Hope language at home, so the family can practice together. Hope is measurable and teachable. With Hopelessness at an all-time high in youth, we've got to be proactive about practicing these skills with youth. This is a tool parents can use to talk about mental health in a positive, proactive way at home and know what to look out for and find resources for support.

The digital guide is available at no cost.
Get your copy here: www.Hopefulminds.org/curriculums

Some may prefer hard copies of our programming. Visit www.Hopefulminds.org/curriculums to find out where to order hard copies for personal use or order for your library or bookstore.

9. Hopeful Minds Deep Dive K-6 Digital Version and Print Version

The Hopeful Minds Deep Dive Educator Guide is a curriculum for K-6 (but can be used for any age) designed to give children a deeper understanding of the core components of Hope. The curriculum is scripted allowing anyone to teach the 16, 45-minute lessons that explore the tools needed to create, maintain, and grow Hope, background information for educators, supplemental resources, classroom visuals, and a Hopework Book for students. It meets National Health Education Standards put forth by the CDC. The Hopework Book is included in the curriculum, but it can also be downloaded separately.

Some may prefer hard copies of our programming. Visit www.Hopefulminds.org/curriculums to find out where to order hard copies for personal use or order for your library or bookstore.

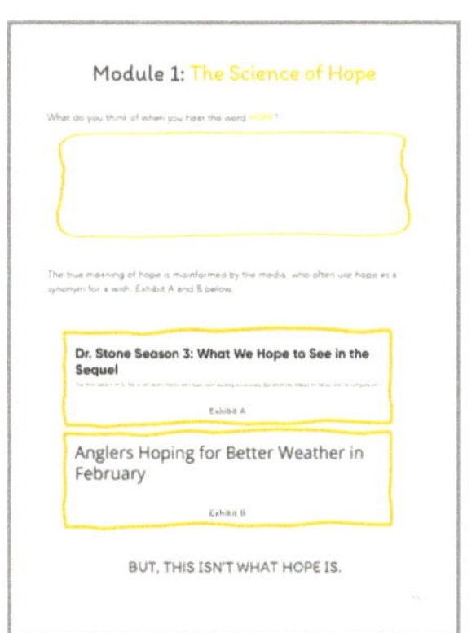

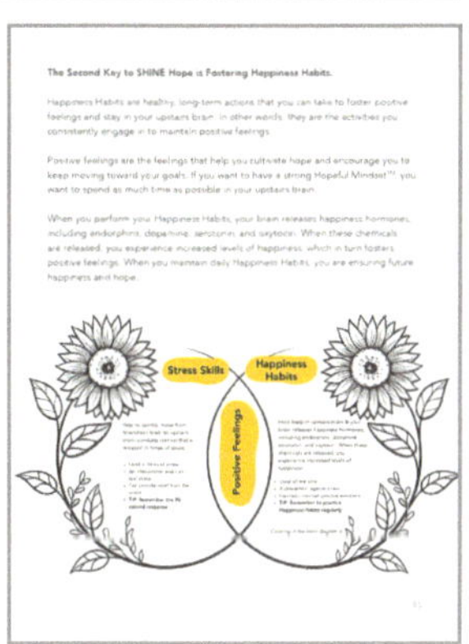

10. Hopeful Minds Teen Hopeguide Digital Version and Print Version

Our Hopeful Minds Teen Hopeful is a 12-module workbook that was tested and approved by teens. The program introduces the Five Keys to Shine Hope™: Stress Skills, Happiness Habits, Inspired Actions, Nourishing Networks, and Eliminating Challenges. This comprehensive approach empowers teens to navigate challenges, embrace positivity, and empower resilience.

The workbook employs a peer-to-peer teaching style, encouraging both individual reflection and group discussions. Interactive elements such as engaging activities, games, creative expressions, worksheets, and puzzles enhance the learning experience. The workbook itself is the curriculum, so educators are not necessary for use. We encourage having facilitators, yet it is designed for anyone to use.

It is based on the award-winning Hopeful Minds program and is intended for teens to do together. As in the US alone, 57% of teen girls are identifying with persistent Hopelessness, we encourage entire schools and classrooms to do the program. It is a proactive, protective approach to mental health. These are skills all youth need to learn and can be implemented in classrooms, after-school programs, places of worship, and more. Anywhere that teens gather, this program is encouraged for use.

Some may prefer hard copies of our programming. Visit www.Hopefulminds.org/curriculums to find out where to order hard copies for personal use or order for your library or bookstore.

11. Hopeful Mindsets Workplace Overview Video Course License for use with Teachers

The Hopeful Mindsets Workplace Overview is a 90-minute video course that introduces Hope and the Five Keys to Shine Hope™ framework to help you create, maintain, and grow Hope in the workplace. We give an overview of the framework, so you can then apply it to your career to activate Hope at work. It gives you the tools you need to practice Stress Skills and Happiness Habits, take Inspired Actions, cultivate Nourishing Networks, and Eliminate Challenges that get in the way of your ability to Hope.

HAPPINESS HABITS: GRATITUDE

This exercise is an excerpt from Dr. Dan Tomasulo's book, "Learned Hopefulness." Take five minutes to write down everything that happened yesterday. Pay attention to what happened, how it made you feel, and how you responded. Focus on your thoughts and feelings during your reflection.

Gratitude is one of the Happiness Habits that can help us stay in our upstairs brain. Take the next five minutes to once again write down everything that happened yesterday, but now, frame it through the lens of gratitude. What were you grateful for? When things didn't go as you wanted or you were triggered, what could you find to still be grateful for?

The course is taught by Kathryn Goetzke, based on her knowledge of mental health, business, Hope, and her work to date, which has been featured around the world and published. It compiles knowledge she learned from her lived experience, and leading experts on Hope, mindset, mental health, stress, positive psychology, business, communications, and more.

The video course is available for individual purchase at www.Hopecourses.com. For bulk license, e-mail us at activate@theshinehopecompany.com.

12. Hopeful Mindsets on the College Campus 10-Module Video Course

Hopeful Mindsets on the College Campus is a 10-module video course from The Shine Hope Company that equips students with crucial Hope skills through expert insights and real-life stories. The course features experts from Harvard, Stanford, and Columbia, with insights from recent college graduates that offer real-life practical strategies and stories from their experiences with homelessness, mental health diagnoses, death, violence, and everyday challenges at school.

The video course is available for individual purchase at www.Hopecourses.com. For bulk license, e-mail us at activate@theshinehopecompany.com.

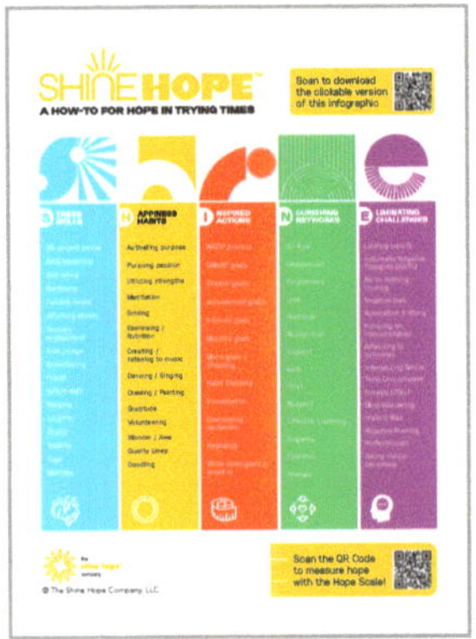

13. Print Shine Hope Educational Moments of Hope Cards, Teen and Adult Brochures, Posters, Signs, and Billboards

Our print communications provide an opportunity to educate all on the Shine Hope framework in an easy-to-understand way. We want to ensure all are equipped with a crisis hotline and skills to proactively activate their Hope. It is our goal to ensure all know that Hope is a skill, measurable, and teachable.

We have two brochures, a general one and a teen-focused one, available for waiting rooms, community centers, lobbies, offices, or libraries. We have posters for the walls in workplaces, coffee shops, libraries, or waiting rooms, and Moments of Hope cards to hand out anywhere people gather. Lastly, we have a Sunflower Fascinating Facts poster for some interesting tidbits on sunflowers, if you have an interest! All of these print materials serve as a cost-effective approach to getting the word out about Hope, and start equipping the population to take a proactive approach to managing Hope in their life and the lives of others.

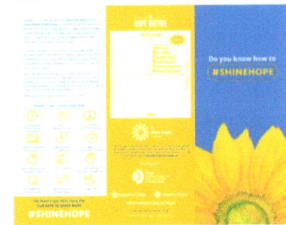

You can purchase printed versions of the Shine Hope Educational materials here: theshineHopestore.com/collections/spread-Hope

See the appendix for billboards, posters, and signs that you can print or purchase to incorporate into your educational campaign about 'how' to Hope.

14. Hope Ambassador Program & Hope Clubs Framework

We are working to build out models for Hopeful Cities Ambassadors and Shine Hope Clubs in schools. In Reno, the city has an Ambassador program where they distribute our Moments of Hope cards, to ensure all know where to go for support. We encourage you to set up an ambassador program in your city.

Be sure to sign-up for our newsletter to hear when we have a Hope Ambassador and Hope Club framework set-up.

15. Community Members Teaching Hope Video Training (police, older population, etc.)

We have expanded our reach by involving police officers in teaching our Hopeful Minds program. Additionally, we offer a video course that provides guidance on implementing similar strategies. With our programs being scripted, anyone proficient in English can effectively teach Hopeful Minds in various settings, including schools, after-school programs, housing authorities, and other community initiatives.

Given that the first key to Shine Hope is identifying and managing the stress response, we believe that engaging police officers in positive and productive interactions with children will enhance community relations. This approach introduces children to the transformative power of Hope while equipping first responders with crucial Hope skills to combat feelings of Hopelessness. We firmly believe that teaching is the best way to learn.

You can find out how to get community members teaching Hope at: www.Hopefulcities.org/community-education/

16. Educator Training

We offer both online and in-person training sessions for educators on the Hopeful Minds program. Our training comprehensively covers the science of Hope and Hopelessness, delves into the Five Keys to Shine Hope, provides insights into available resources for educators, and equips them with effective strategies for engaging youth in the classroom.

Find out more about the training at activate@theshinehopecompany.com.

I love that they are practical and really make hope a science. I have been teaching explicitly about hope in my classroom for almost 15 years and these are the first resources that I have found to support that work.
- Amy B. 4th grade to 9th grade Educator

The children are really enjoying them. Great work in this very important field. *- Lorna, Kindergarten to grade 8th Educator, Gus Wetter School, Canada*

I was struggling to find a curriculum that wasn't a million pages long. Something that children on all levels could grasp and understand. You don't know what a huge help this is to us.
- Woodland Intermediate School 5th Grade Team

Very useful accessible student friendly age-appropriate resources. It has excellent lesson plans.
- Pascal, School Counselor

Our students more than ever need hope in their lives so this program allows us to share hope with them!
- Sara, School Counselor

Read more testimonials at www.hopefulminds.org/hopeful-minds-testimonials/

Workplace

Employers incur $15,000 per year in lost productivity, health care costs, and turnover for every employee who is experiencing emotional distress (NCS, 2023).

Additionally, depression leads to a significant reduction of 11.5 days in productivity every three months, with performance impairment spanning one to two hours within an eight-hour workday (Health Action, 2023). Gallup (2021) recently wrote that ***Hope is one of the four basic needs in strength-based leadership.***

Hopelessness, the primary symptom of depression and a key symptom of anxiety; it is one of the greatest costs to employers. You can use the One Mind at Work calculator to see specific costs to your workplace by visiting: onemindatwork.org/at-work/serious-depression-calculator/

The statistics are clear, higher Hope corresponds to -

- Reduced anxiety and depression (APA, 2013a)
- Increased workplace productivity by 14%, outperforming productivity based on the worker's intelligence, optimism, and self-efficacy (APA, 2013b)
- Increased employee retention (e-Hasan et al., 2022)
- Increased job engagement among individuals who are lonely from remote work (Baraket-Bojmel et al., 2023)
- Increased sleep quality (Trezise et al., 2018)
- Decreased recovery time following an injury.

Hope is associated with setting and achieving goals, as well as being more likely to take action in tough situations, making it a natural fit for workplace training. It also helps employees to see how the work they are doing is meaningful, which in turn can improve happiness and well-being. All of these components create a more successful workplace.

Implementing Hope in the workplace can have a ripple effect that extends beyond the confines of the office, potentially benefiting entire cities. When employees are equipped with Hope, they tend to be more engaged, productive, and mentally resilient. This translates to a workforce that contributes actively to the local economy, stimulates innovation, and fosters a positive work culture.

Engaged and motivated employees are more likely to participate in community initiatives, volunteer programs, and social activities, thus positively impacting the social fabric of the city. Moreover, when companies prioritize employee well-being by nurturing Hope, it sets a precedent for other businesses, creating a collective atmosphere of support, empathy, and growth within the city. This not only improves the quality of life for individuals but also bolsters the overall vitality and spirit of the city itself.

TAKE ACTION WITH THE FOLLOWING INTERVENTIONS

(No cost)

Shine Hope

1. International Day of Hope

The International Day of Hope, scheduled on July 12th, provides an opportunity for global solidarity. On this day, we come together to share the science, stories, and strategies of Hope and actively engage in implementing Hope in our lives and communities worldwide. Our goal is to advocate for the establishment of The International Day of Hope through an official United Nations resolution.

This initiative will kick off a five-day campaign featuring the Five-Day Global Hope Challenge, yard signs, sunflower gardens, workplace educational posters, Shine Hope sunflower murals, live speaking events, classroom teachings of Hopeful Minds, and more.

We encourage Mayors, Governors, and schools to issue proclamations for their towns and environments, demonstrating solidarity in officially recognizing the day and joining the Hopeful Cities Movement.

Day 1
The Impact of Hopelessness, and the Science of Hope
#GlobalHopeChallenge
Be Part Of Activating Hope Around The World

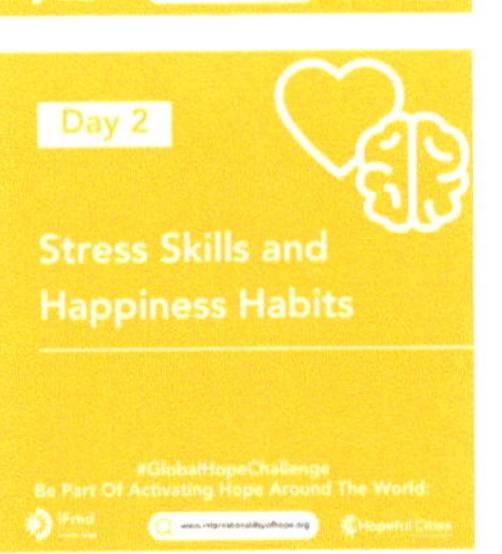
Day 2
Stress Skills and Happiness Habits
#GlobalHopeChallenge
Be Part Of Activating Hope Around The World

2. Five-Day Hope Challenge & Social Media Campaign

The Five-Day Global Hope Challenge is a five-day e-mail challenge ensuring all know the what, why, and how to Hope. It reviews what Hopelessness is, the Shine Hope framework, and instructs how to measure Hope. It is a simple way to get started learning how to Shine Hope.

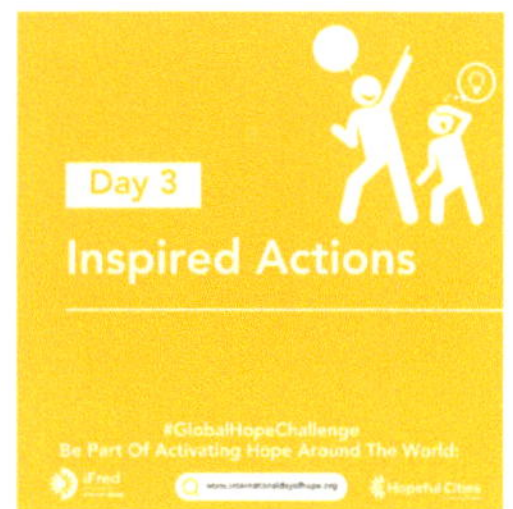
Day 3
Inspired Actions
#GlobalHopeChallenge
Be Part Of Activating Hope Around The World

Social media is a great way to share the resources available for Hope with your friends, family, and community, so we have created a social media toolkit for Hope as well. All of our images and content are available to download at no cost to share and activate the message for Hope. Download the Hope Challenge Social Media Kit for daily social media posts: Hopefulcities.org/social-media/

You can also help your community access the Hopeful Cities resources by tagging us in your posts using @theshineHopecompany @ifredorg #HopefulCities #Hope

3. Digital Communication Assets: Shine Hope Infographic, Moments of Hope Cards, Teen and Adult Brochures, Posters, Signs, and Billboards

SHINE HOPE
A HOW-TO FOR HOPE IN TRYING TIMES

Do you know how to #SHINEHOPE

These are signs for
HOPE

HOPE is Teachable
HOPE is Teachable

Our digital communications provide an opportunity to educate all on the Shine Hope framework in an easy-to-understand way. We have an infographic and digital downloadable tool for newsletters or social media posts to describe the Shine Hope framework at no cost.

We have two brochures that are great for sharing in newsletters or other city-wide communications. Our Digital Shine Hope Brochure is geared toward adults, while our Teen Shine Hope brochure is similar, yet more teen-friendly. Both brochures include tips on the Shine Hope framework, and have resources to help elevate Hope in your life and with others.

We also have Shine Hope Posters you can use as individual images, or on web portals, that give more details into the Shine Hope Framework. Lastly, we have a Sunflower Fascinating Facts poster for some interesting tidbits on sunflowers, if you have an interest!

Get your digital copies here: theshinehopecompany.com/shine-Hope/

See the appendix for billboards, posters, and signs that you can print or purchase to incorporate into your educational campaign about 'how' to Hope.

4. Workplace Research

Encourage all those in your company to measure Hope, so we can start tracking Hopefulness in individuals around the world.

You can find the free scales, and encourage others to take them, here: theshinehopecompany.com/measure-your-Hope/

5. Cause Marketing Campaign

Cause marketing is a great way to activate and educate around Hope. We partner with brands and retailers to license our Shine Hope framework, and work to share the skills and support the work we do in low income areas with the nonprofit we support iFred. If you are a brand, join our mission to teach Hope around the world, and work with us to create a cause marketing campaign about the "how to" of Hope.

If you are interested in working with us on a cause marketing campaign, email us at activate@theshinehopecompany.com

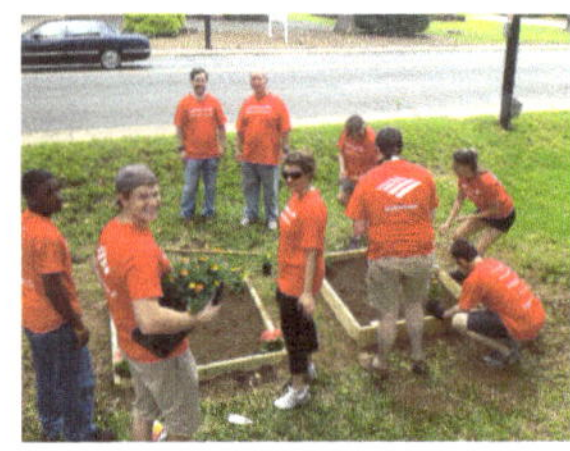

6. Employee Volunteer Opportunities

Our Hopeful Minds programs are all scripted, so your employees are able to work with local schools, after-school programs, youth groups, or other places kids gather to help teach Hope. There are also opportunities to plant gardens or do art projects in low-income areas.

Email us if you want to activate your employees at activate@theshinehopecompany.com.

7. Hopeful Minds Parent's Guide Digital Version

The Hopeful Minds Parent's Guide provides a broad overview of the concepts discussed in the Hopeful Minds curriculums, and provides parents with easy ways to implement the Five Keys to Shine Hope (Stress Skills, Happy Habits, Inspired Actions, Nourishing Networks, and Eliminating Challenges) and Hopeful language in the home. It is helpful when parents reinforce Hope language at Hope, so the family can practice together.

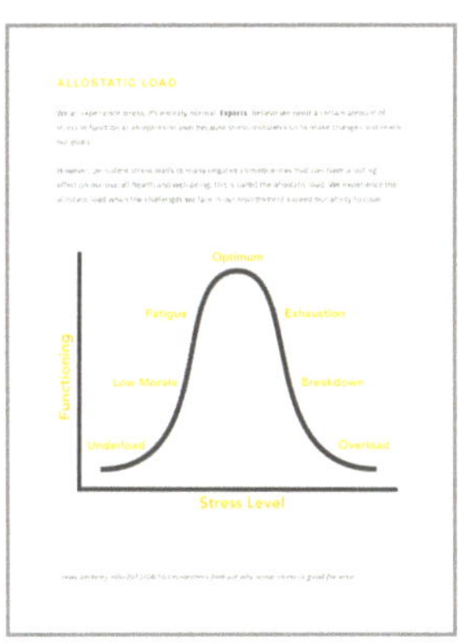

Hope is measurable and teachable. With Hopelessness at an all-time high in youth, we've got to be proactive about practicing these skills with youth. This is a tool parents can use to talk about mental health in a positive, proactive way at home and know what to look out for and find resources for support.

Get your copy here: www.Hopefulminds.org/curriculums

Some may prefer hard copies of our programming. Visit www.Hopefulminds.org/curriculums to find out where to order hard copies for personal use or order for your library or bookstore.

8. Hopeful Minds Overview K-6 Digital Version

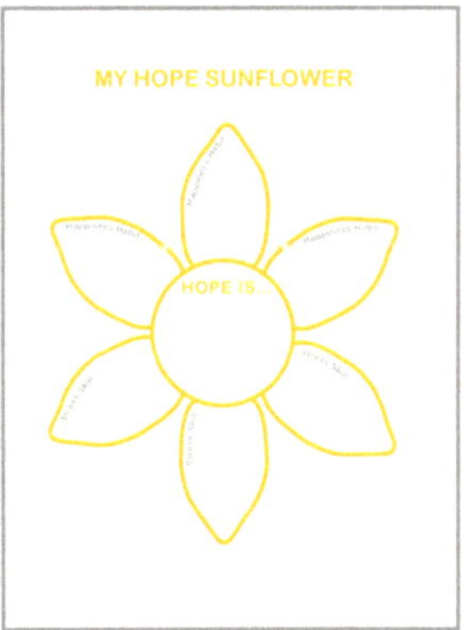

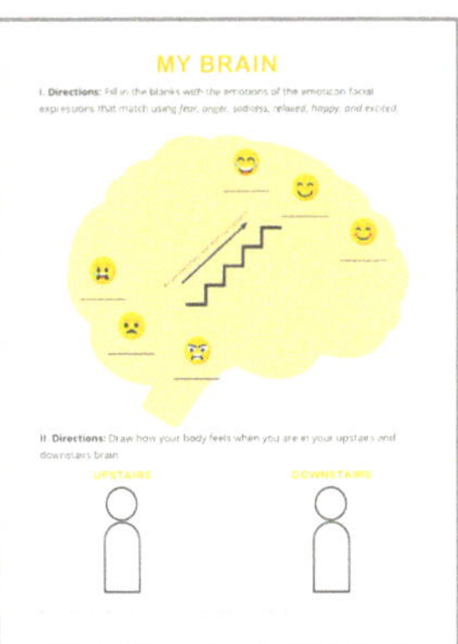

The Hopeful Minds Overview Educator Guide is a curriculum designed to give children an introduction to the "what," "why," and "how" of Hope. The curriculum includes three, one-hour lessons that introduce the key tools needed to create, maintain, and grow Hope. Additionally, the curriculum includes background information for educators, supplemental resources, classroom visuals, and a Hopework Book for students.

While this curriculum is geared towards 2nd-grade students, it can be easily adapted for any age range (adults included) and can be utilized in any setting (such as schools, after-school programs, places of worship, hospitals, offices, and more). This curriculum has been specially designed to be used for either classroom or remote learning.

The digital guide is available at no cost. This is the downloadable version of the curriculum, and we have downloadable fillable workbooks as well.

Get your copy here: www.Hopefulminds.org/curriculums

Some may prefer hard copies of our programming. Visit www.Hopefulminds.org/curriculums to find out where to order hard copies for personal use or order for your library or bookstore.

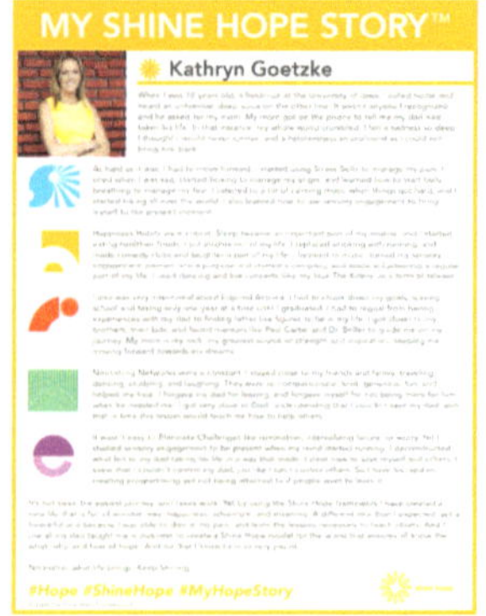

9. My Shine Hope Story™ for Leadership

My Shine Hope Story™ gives community leaders an avenue to normalize Hopelessness and share their unique experiences by using the Shine Hope framework. We all experience moments of Hopelessness (i.e., emotional despair and motivational helplessness). How we manage the moments of Hopelessness matters.

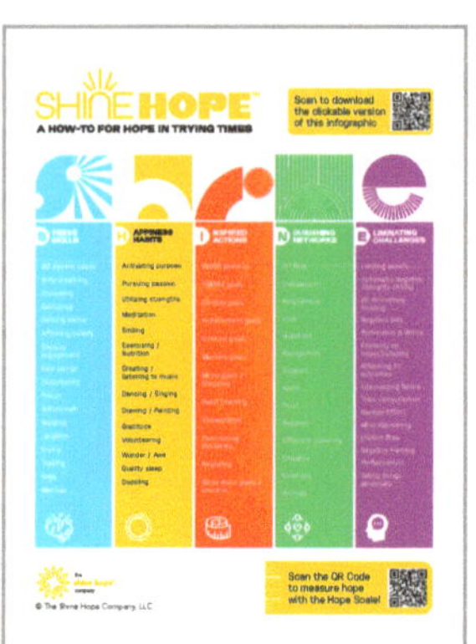

We suggest you share a recent challenge, large or small. We encourage you to spend 20% of the time describing what the Hopelessness was about (sadness, anger, fear, and powerlessness). We then ask you to spend 80% of the story sharing how you overcame it.

What Stress Skills did you use? What Happiness Habits did you practice? What Inspired Actions did you take to make it through? How did your Nourishing Networks support you? And what was one of your biggest challenges (i.e., negative thought pattern), and how did you overcome it?

We created an outline for you to use while creating your Hope Story, and you can use our Shine Hope infographic for some ideas. You can download the Shine Hope Infographic here: theshinehopecompany.com/shine-Hope/

We encourage you to share your story with the community while encouraging others to share their stories as well. Discuss what skills work, help all practice, and be sure people know where to go for support if they are having challenges in your city. You can download the My Shine Hope Story™ Template here: www.Hopefulcities.org/workplace.

If you want to share your Hope story on social media, we encourage you to do so. Please tag us @ifredorg @theshineHopecompany #myHopestory #myHopehero so we can share with others.

(Programs with Fees)

10. Workplace Leadership & Management Training

We offer diverse options for management and leadership training, incorporating the measurement of managers' Hope levels and the use of the VIA Strength Finder—a powerful tool that identifies strengths to optimize overall leadership performance.

This training covers various aspects, including insights into the Hope Matrix, an exploration of the definition and consequences of Hopelessness, and the provision of strategies for proactively managing moments of Hopelessness while instilling the necessary skills to activate Hope.

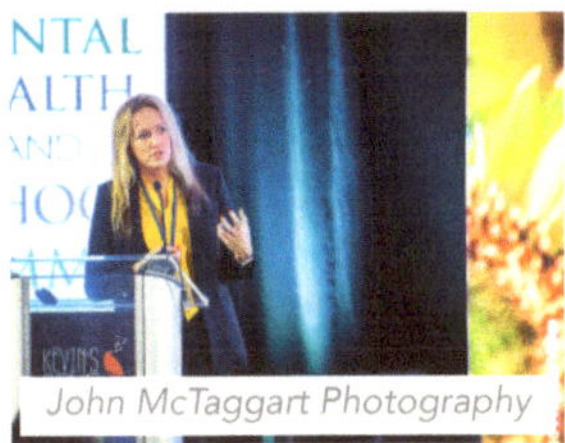

Managers are specifically guided on the effective application of the Shine framework in their leadership roles, integrating our Hopeful Minds Overview for the Workplace workbook. They are encouraged to explore how Hope can serve as inspiration and support for their teams.

The primary goal is to equip company leaders with essential skills and ensure their awareness of available resources before reaching a crisis point. The program is delivered through a 90-minute in-person meeting, fostering a collaborative environment that emphasizes learning and practical application.

For executive leadership, high-level Leadership Training is offered to provide a profound understanding of the science of Hope. This training introduces the Shine framework, offering strategies and techniques to nurture Hope in the workplace. It includes a presentation tailored to industry-specific statistics. Your leadership team will gain insights into the psychology of Hope, measurement methods, industry-specific Hope considerations, cultivating Hopeful mindsets, and the practical application of Hope in leadership.

The executive leadership training session is designed to meet the needs of top leaders and their busy schedules and thus are shorter (i.e., 15- 60 minutes); however, the training can be tailored according to needs.

To complement management and leadership training, the Hopeful Mindsets Workplace Overview video course serves as an excellent addition, ensuring that entire companies understand the what, why, and how of Hope.

To learn more about leadership training, please email us at activate@theshinehopecompany.com.

11. Hopeful Mindsets Workplace Overview Video Course License

The Hopeful Mindsets Overview Workplace Video Course offers a comprehensive understanding of Hope and its practical application in the workplace. The annual license grants access to the 90-minute video course for up to 250 employees, providing them with the tools and knowledge to cultivate Hopeful mindsets and promote a positive work environment. We will also provide access to Shine Hope Posters that can be used to reinforce the Shine framework throughout the workplace.

The video course is available for individual purchase at www.Hopecourses.com. For bulk license, e-mail us at activate@theshinehopecompany.com.

12. Workplace Keynote Presentation at Conferences

A workplace keynote presentation offers the chance to address your entire company or present at a prominent industry conference, delivering a 60-minute session focused on teaching Hope. Our objective is to gauge the level of Hope among employees in real-time and subsequently delve into the essential skills on the what, why, and how of Hope. Participants will gain insights into effectively applying the Shine framework in their respective roles, integrating the workbook associated with the overview video course, and exploring methods to inspire and support their teams.

To request Kathryn Goetzke to keynote, please email us at activate@theshinehopecompany.com

13. Print Shine Hope Communication Assets: Moments of Hope Cards, Teen and Adult Brochures, Posters, Signs, and Billboards

Our print communications provide an opportunity to educate all on the Shine Hope framework in an easy-to-understand way. We want to ensure all are equipped with a crisis hotline and skills to proactively activate their Hope. It is our goal to ensure all know that Hope is a skill, measurable, and teachable.

We have two brochures, a general one and a teen-focused one, available for waiting rooms, community centers, lobbies, offices, or libraries. We have posters for the walls in workplaces, coffee shops, libraries, or waiting rooms, and Moments of Hope cards to hand out anywhere people gather. Lastly, we have a Sunflower Fascinating Facts poster for some interesting tidbits on sunflowers, if you have an interest! All of these print materials serve as a cost-effective approach to getting the word out about Hope, and start equipping the population to take a proactive approach to managing Hope in their life and the lives of others.

You can purchase printed versions of the Shine Hope Educational materials here: theshineHopestore.com/collections/spread-Hope

See the appendix for billboards, posters, and signs that you can print or purchase to incorporate into your educational campaign about 'how' to Hope.

14. Hopeful Minds Parent's Guide Print Version

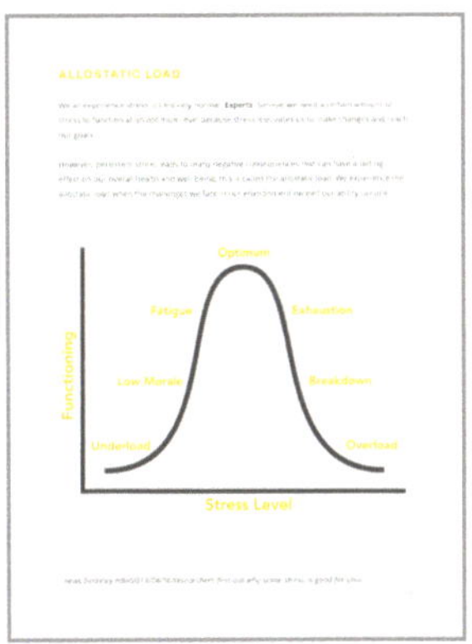

The Hopeful Minds Parent's Guide provides a broad overview of the concepts discussed in the Hopeful Minds curriculums, and provides parents with easy ways to implement the Five Keys to Shine Hope (Stress Skills, Happy Habits, Inspired Actions, Nourishing Networks, and Eliminating Challenges) and Hopeful language in the home. It is helpful when parents reinforce Hope language at home, so the family can practice together. Hope is measurable and teachable. With Hopelessness at an all-time high in youth, we've got to be proactive about practicing these skills with youth. This is a tool parents can use to talk about mental health in a positive, proactive way at home and know what to look out for and find resources for support.

The digital guide is available at no cost.
Get your copy here: www.Hopefulminds.org/curriculums

Some may prefer hard copies of our programming. Visit www.Hopefulminds.org/curriculums to find out where to order hard copies for personal use or order for your library or bookstore.

15. Hopeful Minds Overview K-6 Print Version

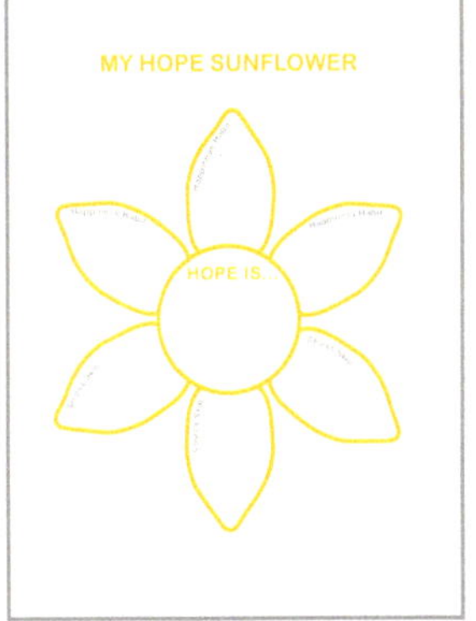

The Hopeful Minds Overview Educator Guide is a curriculum designed to give children an introduction to the "what," "why," and "how" of Hope. The curriculum includes three, one-hour lessons that introduce the key tools needed to create, maintain, and grow Hope. Additionally, the curriculum includes background information for educators, supplemental resources, classroom visuals, and a Hopework Book for students.

While this curriculum is geared towards 2nd-grade students, it can be easily adapted for any age range (adults included) and can be utilized in any setting (such as schools, after-school programs, places of worship, hospitals, offices, and more). This curriculum has been specially designed to be used for either classroom or remote learning.

The digital guide is available at no cost. This is the downloadable version of the curriculum, and we have downloadable fillable workbooks as well.

Get your copy here: www.Hopefulminds.org/curriculums

Some may prefer hard copies of our programming. Visit www.Hopefulminds.org/curriculums to find out where to order hard copies for personal use or order for your library or bookstore.

Healthcare

The CDC (2023) has estimated that 90% of all healthcare costs (i.e., about $3.7 trillion per year) in the United States go towards treating mental health conditions and chronic diseases. Poor health creates a great burden on the economy of a city, as it generates additional costs for employers, may lead to declines in labor force participation, and overtax healthcare providers, often leaving those in marginalized areas undertreated.

Given the associated consequence of an overburdened healthcare system, implementing Hope into cities makes sense. Individuals who are higher in Hope have an-

- Increased optimistic view of life leading them to take action to preserve their health (i.e., more frequent exercising, reduced fat intake, and avoidance of substance use; Berg et al., 2011; Harvard T.H. Chan, 2021; Meraz et al., 2023; Nsamenan & Hirsch, 2014).
- Improved physical health and health-related outcomes, including a reduced risk of all-cause mortality, fewer chronic conditions (i.e., diabetes, hypertension, stroke, cancer, heart disease, lung disease, arthritis, and overweight/obesity, chronic pain; Long et al., 2020)
- Reduced risk of some mental health conditions (i.e., depression, anxiety, and stress; Long et al., 2020)
- Improved sleep patterns (Feldman & Sills, 2013; Long et al., 2020; Senger, 2023)
- Increased sense of connectedness and belonging, which is also linked to prevention and positive health outcomes (Wothington, 2020)

In addition to prevention efforts, Hope is a documented intervention in health outcomes. For example, researchers have found individuals with higher Hope -

- Adhere to treatment plans better than those low in Hope, as they are more motivated to reach recovery goals (e.g., Javanmardifard et al., 2020; Kurita et al., 2020).
- Have expedited recovery times from injuries and diseases, with individuals higher in Hope having a more favorable prognosis for postoperative recovery (Long et al., 2020; Zhu et al., 2017; Zou et al., 2022).

By adopting strategies to enhance healthcare, cities can foster healthier, more inclusive, and resilient communities. Improved health outcomes, reduced healthcare costs, social cohesion, and an overall better quality of life are just some of the benefits that contribute to the city's well-being and prosperity.

TAKE ACTION WITH THE FOLLOWING INTERVENTIONS

(No cost)

1. International Day of Hope

The International Day of Hope, scheduled on July 12th, provides an opportunity for global solidarity. On this day, we come together to share the science, stories, and strategies of Hope and actively engage in implementing Hope in our lives and communities worldwide. Our goal is to advocate for the establishment of The International Day of Hope through an official United Nations resolution.

This initiative will kick off a five-day campaign featuring the Five-Day Global Hope Challenge, yard signs, sunflower gardens, workplace educational posters, Shine Hope sunflower murals, live speaking events, classroom teachings of Hopeful Minds, and more.

We encourage Mayors, Governors, and schools to issue proclamations for their towns and environments, demonstrating solidarity in officially recognizing the day and joining the Hopeful Cities Movement.

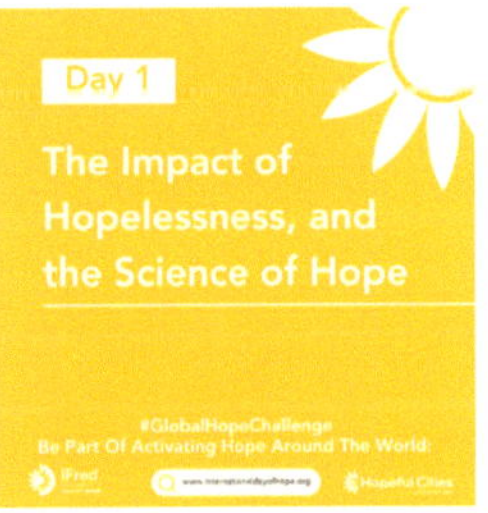

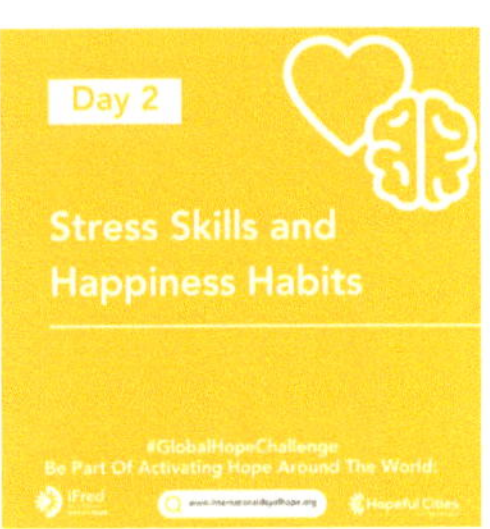

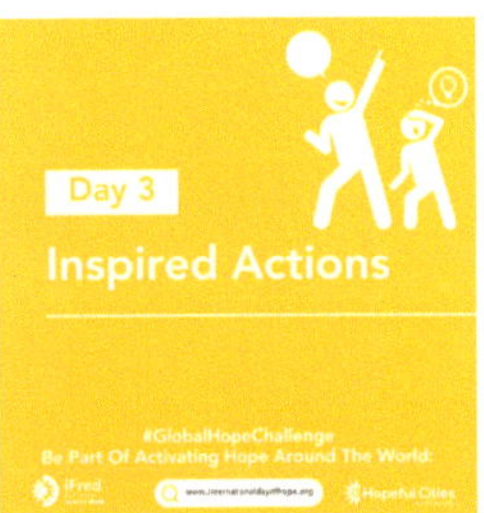

2. Five-Day Hope Challenge & Social Media Campaign

The Five-Day Global Hope Challenge is a five-day e-mail challenge ensuring all know the what, why, and how to Hope. It reviews what Hopelessness is, the Shine Hope framework, and instructs how to measure Hope. It is a simple way to get started learning how to Shine Hope.

Social media is a great way to share the resources available for Hope with your friends, family, and community, so we have created a social media toolkit for Hope as well. All of our images and content are available to download at no cost to share and activate the message for Hope. Download the Hope Challenge Social Media Kit for daily social media posts: Hopefulcities.org/social-media/

You can also help your community access the Hopeful Cities resources by tagging us in your posts using @theshineHopecompany @ifredorg #HopefulCities #Hope

SHINE HOPE
A HOW-TO FOR HOPE IN TRYING TIMES

#SHINEHOPE

These are signs for
HOPE

HOPE is Teachable
HOPE is Teachable

3. Digital Communication Assets: Shine Hope Infographic, Moments of Hope Cards, Teen and Adult Brochures, Posters, Signs, and Billboards

Our digital communications provide an opportunity to educate all on the Shine Hope framework in an easy-to-understand way. We have an infographic and digital downloadable tool for newsletters or social media posts to describe the Shine Hope framework at no cost.

We have two brochures that are great for sharing in newsletters or other city-wide communications. Our Digital Shine Hope Brochure is geared toward adults, while our Teen Shine Hope brochure is similar, yet more teen-friendly. Both brochures include tips on the Shine Hope framework, and have resources to help elevate Hope in your life and with others.

We also have Shine Hope Posters you can use as individual images, or on web portals, that give more details into the Shine Hope Framework. Lastly, we have a Sunflower Fascinating Facts poster for some interesting tidbits on sunflowers, if you have an interest!

Get your digital copies here:
theshinehopecompany.com/shine-Hope/

See the appendix for billboards, posters, and signs that you can print or purchase to incorporate into your educational campaign about 'how' to Hope.

Hopeful Minds
Hopeful Minds Parent's Guide
Planting seeds of HOPE starts at home.

4. Hopeful Minds Parent's Guide Digital Version and Print Version

The Hopeful Minds Parent's Guide provides a broad overview of the concepts discussed in the Hopeful Minds curriculums, and provides parents with easy ways to implement the Five Keys to Shine Hope (Stress Skills, Happy Habits, Inspired Actions, Nourishing Networks, and Eliminating Challenges) and Hopeful language in the home. It is helpful when parents reinforce Hope language at Hope, so the family can practice together.

Hope is measurable and teachable. With Hopelessness at an all-time high in youth, we've got to be proactive about practicing these skills with youth. This is a tool parents can use to talk about mental health in a positive, proactive way at home and know what to look out for and find resources for support.

Some may prefer hard copies of our programming. Visit www.Hopefulminds.org/curriculums to find out where to order hard copies for personal use or order for your library or bookstore.

HAPPINESS HABITS

Directions: Circle your top four favorite happiness habits from the list below. If there are additional happiness habits you use that are not on the list, use the space below the add them.

- Utilizing strengths
- Pursuing passion
- Activating purpose
- Smiling
- Exercising
- Playing or Listening to Music
- Spending time in Nature
- Showing Gratitude and Kindness
- Playing Games
- Volunteering
- Time with Family and Friends
- Experiencing Wonder & Awe
- Practicing Faith
- Sleeping
- Nutrition
- Dancing and Singing
- Donating
- Giving a hug
- Setting Goals
- Practicing Affirmations

My favorite happiness habits:

5. Hopeful Minds Overview K-6 Digital Version and Print Version

The Hopeful Minds Overview Educator Guide is a curriculum designed to give children an introduction to the "what," "why," and "how" of Hope. The curriculum includes three, one-hour lessons that introduce the key tools needed to create, maintain, and grow Hope. Additionally, the curriculum includes background information for educators, supplemental resources, classroom visuals, and a Hopework Book for students.

While this curriculum is geared towards 2nd-grade students, it can be easily adapted for any age range (adults included) and can be utilized in any setting (such as schools, after-school programs, places of worship, hospitals, offices, and more). This curriculum has been specially designed to be used for either classroom or remote learning.

The digital guide is available at no cost. This is the downloadable version of the curriculum, and we have downloadable fillable workbooks as well.

Some may prefer hard copies of our programming. Visit www.Hopefulminds.org/curriculums to find out where to order hard copies for personal use or order for your library or bookstore.

Hope, the 5th Vital Sign of Medicine

-Dr. Edward Barksdale, MD

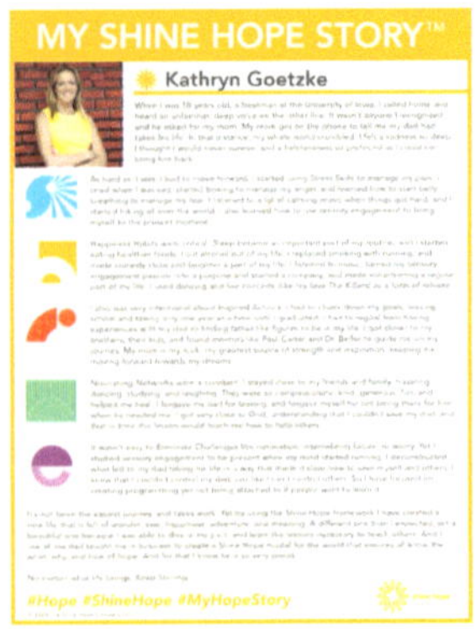

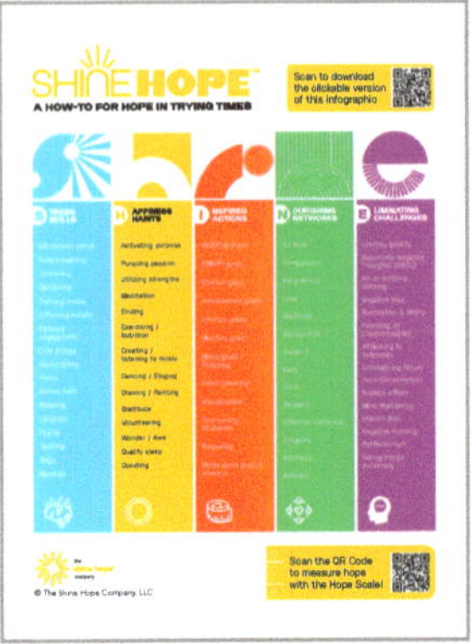

6. My Shine Hope Story™ for Leadership

My Shine Hope Story™ gives community leaders an avenue to normalize Hopelessness and share their unique experiences by using the Shine Hope framework. We all experience moments of Hopelessness (i.e., emotional despair and motivational helplessness). How we manage the moments of Hopelessness matters.

We suggest you share a recent challenge, large or small. We encourage you to spend 20% of the time describing what the Hopelessness was about (sadness, anger, fear, and powerlessness). We then ask you to spend 80% of the story sharing how you overcame it.

What Stress Skills did you use? What Happiness Habits did you practice? What Inspired Actions did you take to make it through? How did your Nourishing Networks support you? And what was one of your biggest challenges (i.e., negative thought pattern), and how did you overcome it?

We created an outline for you to use while creating your Hope Story, and you can use our Shine Hope infographic for some ideas. You can download the Shine Hope Infographic here: theshinehopecompany.com/shine-Hope/

We encourage you to share your story with the community while encouraging others to share their stories as well. Discuss what skills work, help all practice, and be sure people know where to go for support if they are having challenges in your city. You can download the My Shine Hope Story™ Template here: www.Hopefulcities.org/healthcare.

If you want to share your Hope story on social media, we encourage you to do so. Please tag us @ifredorg @theshineHopecompany #myHopestory #myHopehero so we can share with others.

(Programs with Fees)

7. Workplace Leadership & Management Training

We offer diverse options for management and leadership training, incorporating the measurement of managers' Hope levels and the use of the VIA Strength Finder—a powerful tool that identifies strengths to optimize overall leadership performance.

This training covers various aspects, including insights into the Hope Matrix, an exploration of the definition and consequences of Hopelessness, and the provision of strategies for proactively managing moments of Hopelessness while instilling the necessary skills to activate Hope. Managers are specifically guided on the effective application of the Shine framework in their leadership roles, integrating our Hopeful Minds Overview for the Workplace workbook. They are encouraged to explore how Hope can serve as inspiration and support for their teams.

The primary goal is to equip company leaders with essential skills and ensure their awareness of available resources before reaching a crisis point. The program is delivered through a 90-minute in-person meeting, fostering a collaborative environment that emphasizes learning and practical application.

For executive leadership, high-level Leadership Training is offered to provide a profound understanding of the science of Hope. This training introduces the Shine framework, offering strategies and techniques to nurture Hope in the workplace. It includes a presentation tailored to industry-specific statistics. Your leadership team will gain insights into the psychology of Hope, measurement methods, industry-specific Hope considerations, cultivating Hopeful mindsets, and the practical application of Hope in leadership. The executive leadership training session is designed to meet the needs of top leaders and their busy schedules and thus are shorter (i.e., 15- 60 minutes); however, the training can be tailored according to needs.

To complement management and leadership training, the Hopeful Mindsets Workplace Overview video course serves as an excellent addition, ensuring that entire companies understand the what, why, and how of Hope.

To learn more about the management and leadership training, please email us at activate@theshinehopecompany.com

8. Workplace Keynotes

A workplace keynote presentation offers the chance to address your entire healthcare center or present at a prominent industry conference, delivering a 60-minute session focused on teaching Hope. Our objective is to gauge the level of Hope among employees in real-time and subsequently delve into the essential skills about the what, why, and how of Hope. Participants will gain insights into effectively applying the Shine framework in their respective roles, integrating the workbook associated with the overview video course, and exploring methods to inspire and support their teams.

To request Kathryn Goetzke to keynote, please email us at activate@theshinehopecompany.com.

9. Hopeful Mindsets Overview Workplace Video Course License

The Hopeful Mindsets Overview Workplace Video Course offers a comprehensive understanding of Hope and its practical application in the workplace. The annual license grants access to the 90-minute video course for up to 250 employees, providing them with the tools and knowledge to cultivate Hopeful mindsets and promote a positive work environment. We will also provide access to Shine Hope Posters that can be used to reinforce the Shine framework throughout the workplace.

The video course is available for individual purchase at www.Hopecourses.com. For bulk license, e-mail us at activate@theshinehopecompany.com.

We are bringing together a group of really diverse individuals who are working across a lot of different sectors around something that we all can agree is really valuable and important and all communities all around the world, which is Hope.

- Zoya Awan, Director of Public Affairs at Walmart

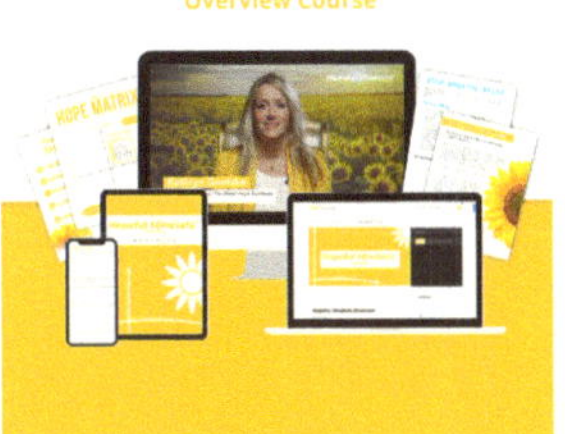

10. Hopeful Mindsets General Overview Video Course License for Patients

The Hopeful Mindsets Overview is a 90-minute video course that introduces Hope and the Five Keys to Shine Hope framework to help you create, maintain, and grow Hope in your life.

This course is taught by Kathryn Goetzke, based on her knowledge of mental health and Hope, and her work to date. It compiles knowledge from leading experts on Hope, Mindset, Mental Health, Stress, Positive Psychology, Business, Communications, and more.

Consider using our General Overview course with patients, for peer-to-peer programs, or for staff.

The video course is available for individual purchase at www.Hopecourses.com. For bulk license, e-mail us at activate@theshinehopecompany.com.

11. Printed Shine Hope Educational Materials for Staff and Waiting Rooms: Moments of Hope Cards, Teen and Adult Brochures, and Posters

Our print communications provide an opportunity to educate all on the Shine Hope framework in an easy-to-understand way. We want to ensure all are equipped with a crisis hotline and skills to proactively activate their Hope. It is our goal to ensure all know that Hope is a skill, measurable, and teachable.

We have two brochures, a general one and a teen-focused one, available for waiting rooms, community centers, lobbies, offices, or libraries. We have posters for the walls in workplaces, coffee shops, libraries, or waiting rooms, and Moments of Hope cards to hand out anywhere people gather. Lastly, we have a Sunflower Fascinating Facts poster for some interesting tidbits on sunflowers, if you have an interest! All of these print materials serve as a cost-effective approach to getting the word out about Hope, and start equipping the population to take a proactive approach to managing Hope in their life and the lives of others.

You can purchase printed versions of our brochures here: theshineHopestore.com/collections/spread-Hope

See the appendix for billboards, posters, and signs that you can print or purchase to incorporate into your educational campaign about 'how' to Hope.

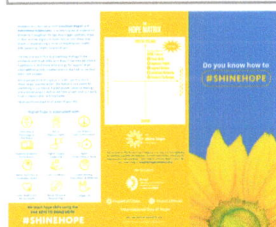

12. Workplace & Patient Research

Consider working with us to do a study specific to your population, positioning you as an innovator in the field. Do a study in your workplace about our Hope interventions and outcomes as it relates to productivity, engagement, absenteeism, health, and sales. Do research with patients to showcase how your interventions are improving their health related outcomes with a focus on ROI.

We love innovating in the field, and have the right Hope experts on hand to help design a study that works for you. Conducting research in Hope science with your company or patients sets you apart as an innovative thought leader. Let's learn and publish more together. Contact us at activate@theshinehopecompany.com to find out more.

13. Online Volunteer Training

Our Hopeful Minds programs are all scripted, so your employees are able to work with local schools, after-school programs, youth groups, or other places kids gather to help teach Hope. There are also opportunities to plant gardens or do art projects in low-income areas.

Email us if you want to activate your employees at activate@theshinehopecompany.com.

Hope is one of the single best predictors of wellbeing.

-Dr. Chan Hellman, PhD

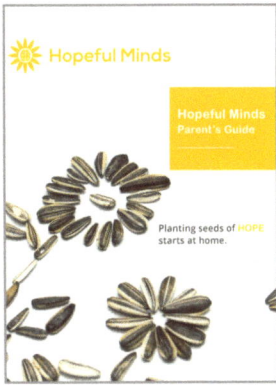

14. Hopeful Minds Parent's Guide Print Version

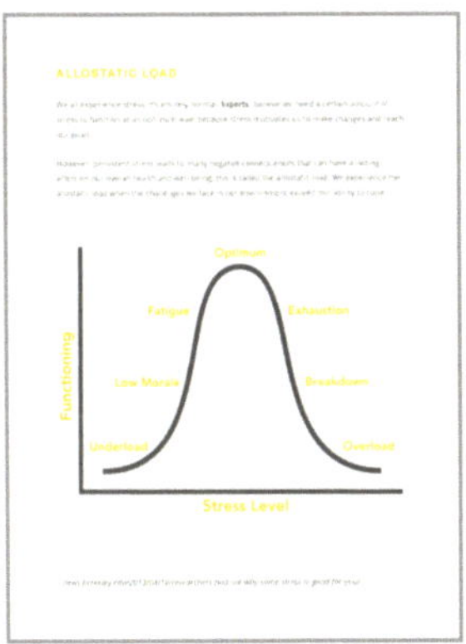

The Hopeful Minds Parent's Guide provides a broad overview of the concepts discussed in the Hopeful Minds curriculums, and provides parents with easy ways to implement the Five Keys to Shine Hope (Stress Skills, Happy Habits, Inspired Actions, Nourishing Networks, and Eliminating Challenges) and Hopeful language in the home. It is helpful when parents reinforce Hope language at home, so the family can practice together. Hope is measurable and teachable. With Hopelessness at an all-time high in youth, we've got to be proactive about practicing these skills with youth. This is a tool parents can use to talk about mental health in a positive, proactive way at home and know what to look out for and find resources for support.

The digital guide is available at no cost.
Get your copy here: www.Hopefulminds.org/curriculums

Some may prefer hard copies of our programming. Visit www.Hopefulminds.org/curriculums to find out where to order hard copies for personal use or order for your library or bookstore.

15. Hopeful Minds Overview K-6 Print Version

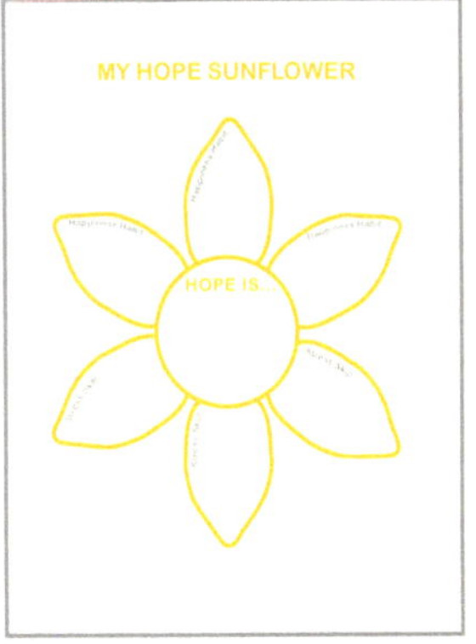

The Hopeful Minds Overview Educator Guide is a curriculum designed to give children an introduction to the "what," "why," and "how" of Hope. The curriculum includes three, one-hour lessons that introduce the key tools needed to create, maintain, and grow Hope. Additionally, the curriculum includes background information for educators, supplemental resources, classroom visuals, and a Hopework Book for students.

While this curriculum is geared towards 2nd-grade students, it can be easily adapted for any age range (adults included) and can be utilized in any setting (such as schools, after-school programs, places of worship, hospitals, offices, and more). This curriculum has been specially designed to be used for either classroom or remote learning.

The digital guide is available at no cost. This is the downloadable version of the curriculum, and we have downloadable fillable workbooks as well.

Get your copy here: www.Hopefulminds.org/curriculums

Some may prefer hard copies of our programming. Visit www.Hopefulminds.org/curriculums to find out where to order hard copies for personal use or order for your library or bookstore.

Art

Nothing is more symbolic than art. Murals, sculptures, photography contests, drawings, and iconic symbols of Hope, with a sign referencing Hopeful Cities at the bottom, serve a dual purpose. They provide beauty in the environment while sharing resources for the "how-to" of Hope in the community.

We aim to not just inspire with our work, but also to teach. To empower both youth and adults around the world with the skills they need to find, create, maintain, and grow Hope. By creating your own artwork for Hope, you are giving us the incredible gift of creatively sharing the message of Hope in your community.

We want people to not just see symbols of Hope, but to be reminded of what they learned about the "how-to" of Hope. We just ask that you use the color yellow, and if so inspired, the sunflower, and add a creative message to a plaque or sign that provides a message about Hope, such as:

DID YOU KNOW? HOPE IS TEACHABLE.
Learn the "how-to" of Hope
at www.Hopefulcities.org

It's important to spread Hope, as Hope isn't just about the self; it is about supporting others on their Hope journeys and sharing strategies for the "how-to" of Hope. When you share your murals, make sure you tag @ifredorg @theshineHopecompany.

Learn more at www.Hopefulcities.org/art

TAKE ACTION WITH THE FOLLOWING INTERVENTIONS

(No cost)

1. International Day of Hope

The International Day of Hope, scheduled on July 12th, provides an opportunity for global solidarity. On this day, we come together to share the science, stories, and strategies of Hope and actively engage in implementing Hope in our lives and communities worldwide. Our goal is to advocate for the establishment of The International Day of Hope through an official United Nations resolution.

This initiative will kick off a five-day campaign featuring the Five-Day Global Hope Challenge, yard signs, sunflower gardens, workplace educational posters, Shine Hope sunflower murals, live speaking events, classroom teachings of Hopeful Minds, and more.

We encourage Mayors, Governors, and schools to issue proclamations for their towns and environments, demonstrating solidarity in officially recognizing the day and joining the Hopeful Cities Movement.

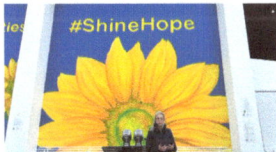

2. Shine Hope Sunflower Mural Templates

The sunflower mural provides an opportunity for community participation in creating large-scale artwork that symbolizes unity and Hope. Examples of sunflower murals are available in our Appendix.

A mural is a piece of artwork that is implemented on a wall. The best placement for a mural that promotes Hope is somewhere public, where it will inspire as many people as possible. It is important to use the mural to teach Hope, not just create art.

We worked with Hayley Meadows of Reno, Nevada to create four murals that you can use in your community. Feel free to use ours or come up with your own. We include an informational plaque with all of our murals so that people are able to visit our website and receive the resources. We never know who we might touch through the power of art, and Hope.

Encourage community memes to take pictures with the art installations and share specific skills on how they Shine Hope and share on social media. @ifredorg and @theshineHopecompany using the following hashtags:

#HopefulCities #SpreadHope #GrowHope #ShineHope

#GrowHope represents our ability to move from Hopeless to Hope, no matter what life brings. Like the sunflower rising from the building, we can rise from adversity with Hope.

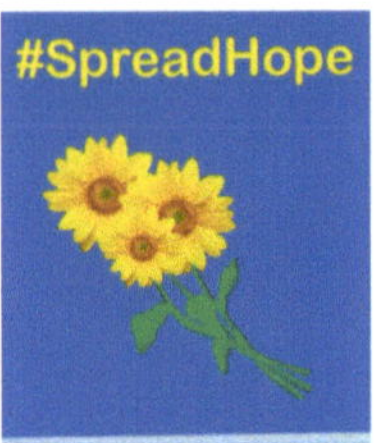

#SpreadHope is a bouquet of sunflowers that people can pretend to hold. When people share photos holding the sunflowers, we encourage them to also share how they #SpreadHope.

#HopefulCities is about the power of community. The field of flowers shows us how to stand all together for Hope and work together to activate Hope in our community.

#ShineHope is a mural with a massive sunflower that people can stand in front of and share how they Shine Hope. Share your favorite Shine skills (Stress Skills, Happiness Habits, Inspired Actions, Nourishing Networks, and Eliminating Challenges) and how they help you #ShineHope.

3. Sunflower Fascinating Facts Digital Poster

Sunflowers are used as the symbol for Hope so we like to incorporate sunflowers as we go. Sunflowers. Our Sunflower Fascinating Facts poster (see Appendix) provides some interesting tidbits on sunflowers to keep you engaged, inspired, and reminded about why we chose such an iconic image.

Download the poster at theshinehopecompany.com/shine-Hope/

4. Sunflower Gardens for Hope

We encourage you to plant sunflower garden in your yard, at your school, or start a community garden of sunflowers as a way to activate Hope. Planting sunflower gardens is good for your health. Sunflower seeds are beneficial for brain health. Bright sunflowers spread happiness. Check out our Hope Marketing Tools page to download a free sign to let people know why you've planted your garden, and where to go for support as the website is right on the sign. A sample garden sign is also included in the Appendix and available for download. You can bring to your printer, and have them add your logo if so inspired to customize for your community.

Learn more at Hopefulcities.org/print-campaign

5. Sunflower Fields for Hope

We planted our first sunflower field for Hope in Ghana over 15 years ago as a way to start the conversation about depression and Hope. Sunflower fields are iconic, gorgeous, and make an even bigger statement about Hope. They help spread the message that Hope is teachable.

If you add a billboard or large sign to the field with a link to the website, it is another great way to send people to Hopeful Cities, and help them activate Hope and get valuable Hope resources. You can dedicate your current sunflower field to Hope, or reach out to local farmers and ask to have a sunflower field dedicated to the cause.

Download the signs at Hopefulcities.org/print-campaign/.

6. Public Art Hope Health Campaigns

Infuse sunflowers into the art around the city to spread awareness about Hope and activate your community. Sculptures are beautiful and creative ways to generate artwork that will last all year long. Create a sunflower sculpture or an iconic art installation using any material you want, share the "Shine" acronym, and include a placard about the importance of Hope to help people start activating Hope and finding resources.

Invite community participation in creating large-scale artwork that symbolizes unity and Hope; find examples in our Appendix. We have created all kinds of posters, templates, and signs to drive awareness about the advantage of Hope, as you want to be sure to educate while you inspire. The opportunities are as endless as your imaginations.

Share what you create for Hope on social media using **@theshineHopecompany** and **@ifredorg #HopefulCities #SpreadHope #GrowHope #ShineHope**

See the appendix for billboards, posters, and signs that you can print or purchase to incorporate into your educational campaign about 'how' to Hope. Make sure to encourage all to practice and learn skills and drive them to our website so they know where they can get support.

(Programs with Fees)

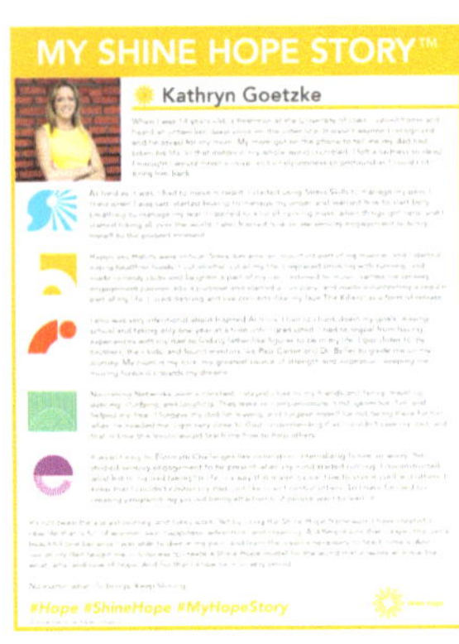

7. Hope Heroes Exhibit

It is important to showcase how people in the community activate the Shine Hope framework, so we learn from each other. The goal of this exhibit is to amplify challenges people might face, and even more importantly the skills and strategies used to overcome the challenge in healthy ways.

We highly encourage you doing a youth version, so youth can learn from others.

Exhibit:

- Have people measure their Hope at the beginning of the exhibit with the statement: Did you know? Hope is measurable (and teachable).
- Share the Shine framework, showcasing how we practice skills for Hope.
- Have a visual showcase of individual Hope Stories throughout the exhibit.
    - Have individuals write a paragraph sharing how they use the Shine framework or use a 'My Shine Hope Story™' template provided.
    - Have a local photographer take photos of the individuals
    - Put the photos and Hope stories on a wall, showcasing the many challenges we face yet as importantly what skills we might use to overcome them.
- Have a Shine Hope sunflower temporary mural on a wall and encourage visitors to take their own photo / share their own Hope story on social media **@theshineHopecompany** and **@ifredorg #HopefulCities #SpreadHope #GrowHope #ShineHope**

8. Interactive Art Educational Experiences

Consider doing an exhibit at an art institute or children's museum. We are happy to work with you to come up with innovative ways to start teaching all about Hope through art. In the past we have done photography contests and art programs, encouraging budding photographers to encapsulate the power of Hope.

Email us at activate@theshinehopecompany.com to find out about how to order this iconic installation for your city.

9. Iconic Large Hope Letters

We have a fantastic artist in Reno, NV that is available to make Hope letters for your community. Iconic, innovative, yellow, lit, metal HOPE letters to stand as a focal point for activating your city.

Pricing starts at 100k. Email us at activate@theshinehopecompany.com to find out about how to order this iconic installation for your city.

Appendix

Branding Guidelines

OUR LOGO

FULL COLOR

WHITE

Iconography

Branding Guidelines

TYPEFACE

NEXA BOLD

ABCDEFGHIJKLMNOPQRSTUVWXYZ
abcdefghijklmnopqrstuvwxyz
1234567890

MONTSERRAT REGULAR

ABCDEFGHIJKLMNOPQRSTUVWXYZ
abcdefghijklmnopqrstuvwxyz
1234567890

COLOUR PALETTE

	YELLOW	GREY	BLACK
HEX	#FFC72C	#97999B	#000000
CMYK	0% 23% 93% 0%	45% 35% 36% 1%	0% 0% 0% 100%

HOPEFUL CITIES BRAND GUIDELINES | Typeface | Colour Palette VOL. 03 OCTOBER 2023

Branding Guidelines

DESIGN GUIDELINES

DO NOT USE ARTIFICIAL SHADOWS WHEN DESIGNING FOR OUR MARKETING MATERIALS. USE BRAND COLORS INSTEAD.

DO NOT STRETCH, ALTER, DISTORT OR ANY OTHER REINTERPRETATION OF LOGOS.

Branding Guidelines

DESIGN GUIDELINES

IMAGES

USE MATERIALS THAT ARE FREE TO USE COMMERCIALLY. YOU MAY DOWNLOAD FREE STOCK PHOTOS IN BIGSTOCKPHOTO.COM OR OTHER SITES THAT OFFERS FREE DOWNLOADABLE STOCK VIDEOS AND IMAGES.

YOU CAN VISIT:
- O-DAN.NET
- PEXELS.COM
- PIXABAY.COM

YOU ARE ALSO ENCOURAGED TO USE OUR OWN PHOTOS FOR MARKETING MATERIALS TO PROMOTE HOPE.

YOU MUST GIVE ATTTRIBUTION IF YOU USE OUR IMAGES.

WE ENCOURAGE THE USE OF POSITIVE, HOPEFUL IMAGES INCLUDING THE COLOR YELLOW AND SUNFLOWERS.

Branding Guidelines

DESIGN GUIDELINES

POSTERS

OUR HOPEFUL CITIES POSTERS HAVE BEEN CAREFULLY CREATED TO SHARE THE MESSAGE OF HOPE IN YOUR COMMUNITY. LEARN MORE AT HOPEFULCITIES.ORG/CITIES

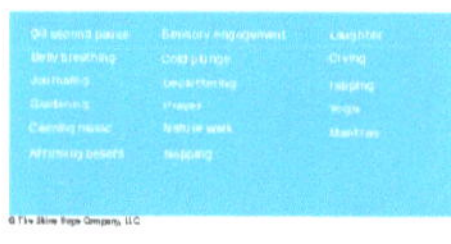

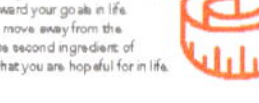

Branding Guidelines

DESIGN GUIDELINES

BILLBOARDS

WE HAVE CREATED BILLBOARD IMAGES IN BOTH ENGLISH AND SPANISH SO THAT YOU CAN HELP SPREAD THE MESSAGE OF HOPE AROUND THE WORLD: HOPEFULCITIES.ORG/CITIES

HOPE is Teachable

Resources available at www.hopefulcities.org

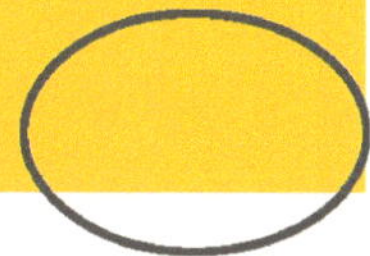

HOPE is Teachable

Resources available at www.hopefulcities.org

MULTIPLE FORMATS AVAILABLE TO ALLOW YOU TO ADD YOUR OWN SPONSORSHIP LOGOS.

Branding Guidelines

DESIGN GUIDELINES

YARD SIGNS

YARD SIGNS ACTIVATE HOPE IN YOUR COMMUNITY, BY PROMOTING THE HOPEFUL CITIES WEBSITES TO GET FOLKS ACCESS TO FREE RESOURCES TO LEARN HOPE SKILLS. LEARN MORE AT HOPEFULCITIES.ORG/CITIES

HOPE
is teachable
#HopefulCities
www.hopefulcities.org

HOPE
is teachable
#HopefulCities
www.hopefulcities.org

MULTIPLE FORMATS AVAILABLE TO ALLOW YOU TO ADD YOUR OWN SPONSORSHIP LOGOS.

Branding Guidelines

DESIGN GUIDELINES
SOCIAL MEDIA POSTS

USE SHORT SENTENCES OR SLOGANS ONLY TO DESCRIBE THE SUBJECT. MAKE SURE TO INCLUDE HASHTAGS. FOCUS ON HOPE AND SUCCESSFUL TREATMENTS AND STRATEGIES FOR COPING. WE HAVE ALSO PROVIDED SOCIAL MEDIA KITS FOR DOWNLOAD.

NO SPONSOR LOGO

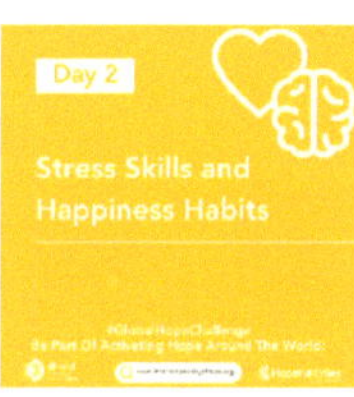

WITH SPONSOR'S LOGO

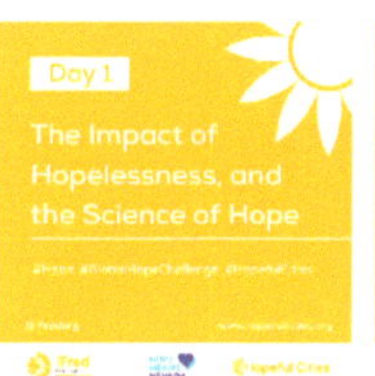

Branding Guidelines

SOCIAL MEDIA GUIDELINES

SOCIAL MEDIA MESSAGING

THE MISSION OF INTERNATIONAL FOUNDATION FOR RESEARCH AND EDUCATION ON DEPRESSION (IFRED) IS TO SHINE A POSITIVE LIGHT ON MENTAL HEALTH AND ELIMINATE THE STIGMA THROUGH PREVENTION, RESEARCH AND EDUCATION.

THE MISSION OF ALL OF OUR SOCIAL MEDIA PLATFORM IS TO LISTEN, ENGAGE, INFORM, AMPLIFY, EDUCATE AND INSPIRE ABOUT MENTAL HEALTH AND HOPE USING REPUTABLE SOURCES AND INSPIRING IDEAS.

TARGET AUDIENCE

CAUSE ADVOCATES
MEDICAL PROFESSIONALS
CURRENT AND POTENTIAL PARTNERS
- CELEBRITIES
- CORPORATIONS
- MEDIA

INDIVIDUALS, BUSINESSES, AND SCHOOLS INTERESTED IN ACTIVATING HOPE

POSTING DOs AND DONTs

DO:

POST POSITIVE ARTICLES, RELEVANT NEW RESEARCH (PROVEN VIA CLINICAL TRIALS OR OTHER ESTEEMED RESOURCES), CELEBRITY MENTAL HEALTH STORIES, AND SCIENCE, STORIES, AND STRATEGIES OF HOPE

FOCUS ON HOPE AND SUCCESSFUL TREATMENTS AND STRATEGIES FOR COPING.

UTILIZE THE WORD "MENTAL HEALTH" IN PLACE OF "MENTAL ILLNESS."

ENGAGE WITH OTHERS. POST COMMENTS, RESPOND TO HASHTAGS, GIVE ENCOURAGEMENT, RETWEET, TAG, AND REPOST.

DO NOT:

PROVIDE MEDICAL ADVICE.

Branding Guidelines

SOCIAL MEDIA GUIDELINES

SOCIAL MEDIA HASHTAGS

WE BELIEVE HOPE IS A UNIVERSAL RIGHT, AND WE AIM TO SHARE HOPE AND HOPE SKILLS WITH EVERYONE. BY USING THESE HASHTAGS, YOU HELP US TO UTILIZE CONSISTENT LANGUAGE AND BRANDING TO ACTIVATE HOPE ON A GLOBAL SCALE.

HOPEFUL CITIES HASHTAGS

#HopefulCities
#GlobalHopeChallenge
#HopeInTheWorkplace
#GrowHope
#SpreadHope
#ShineHope

GENERAL HASHTAGS`

#Hope
#Depression
#Anxiety
#MentalHealth
#YMentalHealth
#YouthMentalHealth

IFRED HASHTAGS

#HopeIsTeachable
#IChooseHope
#ShareHope
#HopefulMinds
#TeachHope
#HopeScience
#HopeResearch
#HopeStory

SOCIAL MEDIA HANDLES

WEBSITE
www.ifred.org

FACEBOOK
@ifredorg / @theshinehopecompany
www.facebook.com/ifredorg
www.facebook.com/theshinehopecompany

INSTAGRAM
@ifredorg / @theshinehopecompany
www.instagram.com/ifredorg
www.instagram.com/theshinehopecompany

LINKEDIN
iFred / The Shine Hope Company
https://www.linkedin.com/company/ifredorg
https://www.linkedin.com/company/theshinehopecompany

TWITTER
@iFredorg / @theshinehopeco
www.twitter.com/ifredorg
www.twitter.com/theshinehopeco

PINTEREST
IFred
www.pinterest.ph/ifredorg

YOUTUBE
ifredorg / @theshinehopecompany
www.youtube.com/@ifredorg
www.youtube.com/@theshinehopecompany

HOPEFUL CITIES BRAND GUIDELINES | Social Media Handles VOL. 03 OCTOBER 2023

Branding Guidelines

Branding Guidelines

DID YOU KNOW
You can learn HOPE for FREE?
#GlobalHopeChallenge
Take the challenge at:

It's time to choose
HOPE
Be Part Of Activating Hope Around The World:

5 days. 5 lessons. 5 actions.
The Five Day Global Hope Challenge starts tomorrow!
SIGN UP NOW!
#GlobalHopeChallenge
Take the challenge at:

Today, we celebrate the
International Day of Hope

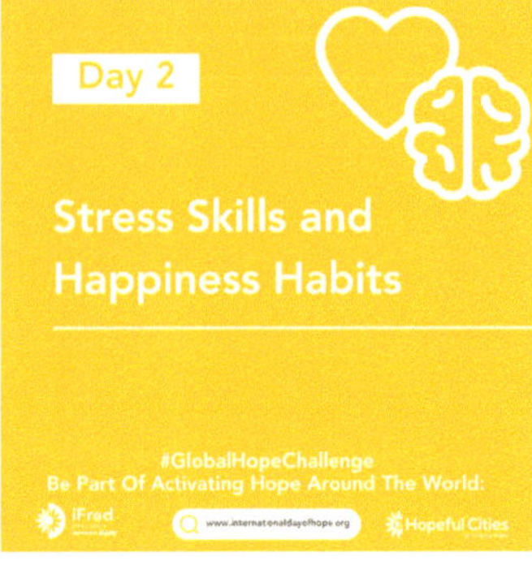

Day 4
Nourishing Networks
#GlobalHopeChallenge
Be Part Of Activating Hope Around The World:

Day 5
Eliminating Challenges
#GlobalHopeChallenge
Be Part Of Activating Hope Around The World:

Day 5
I completed the Five Day #GlobalHopeChallenge!
#GlobalHopeChallenge
Be Part Of Activating Hope Around The World:
www.hopefulcities.org

Social Media Toolkit

The Five-Day Global Hope Challenge is a five-day e-mail challenge ensuring all know the what, why, and how to Hope. It reviews what Hopelessness is, the Shine Hope framework, and instructs how to measure Hope. It is a simple way to get started learning how to Shine Hope.

Social media is a great way to share the resources available for Hope with your friends, family, and community, so we have created a social media toolkit for Hope as well. All of our images and content are available to download at no cost to share and activate the message for Hope. Download the Hope Challenge Social Media Kit for daily social media posts:
Hopefulcities.org/social-media/

You can also help your community access the Hopeful Cities resources by tagging us in your posts using @theshineHopecompany @ifredorg #HopefulCities #Hope

The Challenge covers the following topics:

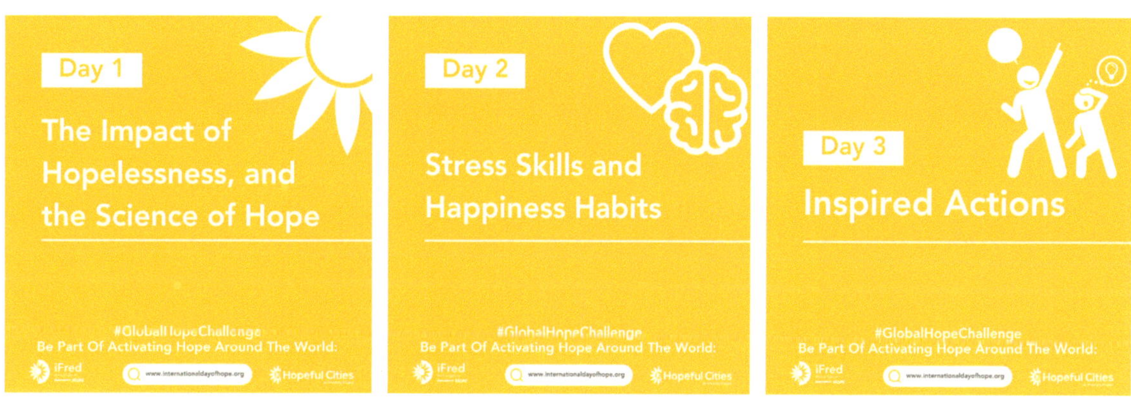

Five-Day Global Hope Challenge

MY SHINE HOPE STORY™

HOW HOPEFUL ARE YOU?

Did you measure your hope? The lower your score, the more you want to practice these skills! Remember, hope is a muscle we need to build it (add it).

Check out here to get your hope score.

To write your own shine hope story, spend 20% of your time writing about your challenge, and 80% of the time sharing strategies for how you overcame it so others can learn from you. Here's how:

1. Write your name in the yellow line next to the box (feel free to use a nickname or anything else).
2. Put your favorite photo on the yellow box, or an image of something that represents you.
3. Write an introduction to your story explaining the challenge you faced. Explain the two ingredients of hopelessness: despair (feelings) and helplessness (inability to act) you experienced.
4. Share sadness, anger, fear, or other feelings, and choose **3 Stress Skills** you used to naviate them (from the Shine infographic, or choose your own!).
5. Share **3 Happiness Habits** you used to get back to your upstairs brain.
6. Talk about **3 Inspired Actions** you took, or share how you chunked down goals, the types of goals you set, or if you had to regoal.
7. Share who was in your **Nourishing Network**, and how they helped you navigate the challenge.
8. Pick 3 challenges from the **'Eliminating Challenges'** on the infographic, and share how you eliminated them.
9. Write your conclusion. What do you want the world to know? What do you wish someone had told you? What is the moral of the story?

If you're inspired, share your story so we can help activate these skills globally.

#Hope #ShineHope #MyShineHopeStory

We all experience moments of hopelessness (emotional despair and motivational helplessness). The key is to use the Shine Hope skills to navigate your way from despair to positive feelings, and helplessness to inspired actions. Use the Shine Hope framework to build your muscle.

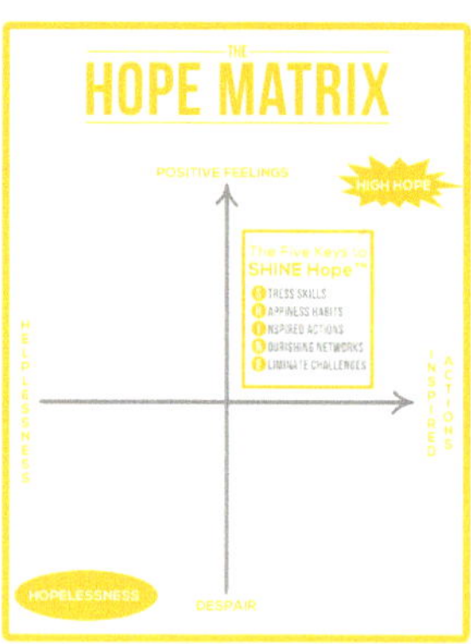

My Shine Hope Story™ Template

MY SHINE HOPE STORY™

Kathryn Goetzke

When I was 18 years old, a freshman at the University of Iowa, I called home and heard an unfamiliar, deep voice on the other line. It wasn't anyone I recognized, and he asked for my mom. My mom got on the phone to tell me my dad had taken his life. In that instance, my whole world crumbled. I felt a sadness so deep I thought I would never survive, and a helplessness so profound as I could not bring him back.

As hard as it was, I had to move forward. I started using Stress Skills to manage my pain. I cried when I was sad, started boxing to manage my anger, and learned how to start belly breathing to manage my fear. I listened to a lot of calming music when things got hard, and I started hiking all over the world. I also learned how to use sensory engagement to bring myself to the present moment.

Happiness Habits were critical. Sleep became an important part of my routine, and I started eating healthier foods. I cut alcohol out of my life. I replaced smoking with running, and made comedy clubs and laughter a part of my life. I listened to music, turned my sensory engagement passion into a purpose and started a company, and made volunteering a regular part of my life. I used dancing and live concerts (like my fave The Killers) as a form of release.

I also was very intentional about Inspired Actions. I had to chunk down my goals, leaving school and taking only one year at a time until I graduated. I had to regoal from having experiences with my dad to finding father-like figures to be in my life. I got closer to my brothers, their kids, and found mentors like Paul Carter and Dr. Belfer to guide me on my journey. My mom is my rock, my greatest source of strength and inspiration, keeping me moving forward towards my dreams.

Nourishing Networks were a constant. I stayed close to my friends and family, traveling, dancing, studying, and laughing. They were so compassionate, kind, generous, fun, and helped me heal. I forgave my dad for leaving, and forgave myself for not being there for him when he needed me. I got very close to God, understanding that I couldn't save my dad, and that in time this lesson would teach me how to help others.

It wasn't easy to Eliminate Challenges like rumination, internalizing failure, or worry. Yet I studied sensory engagement to be present when my mind started running. I deconstructed what led to my dad taking his life in a way that made it clear how to save myself and others. I knew that I couldn't control my dad, just like I can't control others. So I have focused on creating programming yet not being attached to if people want to learn it.

It's not been the easiest journey, and takes work. Yet by using the Shine Hope framework I have created a new life that is full of wonder, awe, happiness, adventure, and meaning. A different one than I expected, yet a beautiful one because I was able to dive in my pain, and learn the lessons necessary to teach others. And I use all my dad taught me in business to create a Shine Hope model for the world that ensures all know the what, why, and how of hope. And for that I know he is so very proud.

No matter what life brings, Keep Shining.

#Hope #ShineHope #MyHopeStory

My Shine Hope Story™ Template

MY SHINE HOPE STORY™

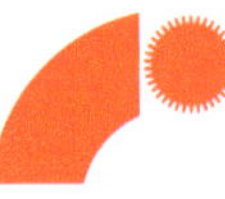

#Hope #ShineHope #MyHopeStory

My Shine Hope Story™ Template

MY HOPE HERO

HOW HOPEFUL ARE YOU?

Did you measure your hope? The lower your score, the more you want to practice these skills! Remember, hope is a muscle we need to build it (add it).

Check out here to get your hope score.

To write your hope hero journey, spend 20% of your time writing about their challenge, and 80% of the time sharing strategies for how they overcame it so others can learn from it. Here's how:

1. Write your hope hero's name in the yellow line next to the box (feel free to use a nickname or anything else).

2. Put your favorite photo of them on the yellow box, or an image of something that represents your hope hero.

3. Write an introduction explaining the challenge they faced. Explain the two ingredients of hopelessness: despair (feelings) and helplessness (inability to act) they experienced.

4. Share sadness, anger, fear, or other feelings, and choose 3 **Stress Skills** they used to navigate them (from the Shine infographic, or choose your own!).

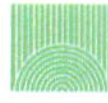
5. Share 3 **Happiness Habits** they used to get back to upstairs brain.

6. Talk about 3 **Inspired Actions** they took, or share how your hope hero chunked down goals, the types of goals they've set, or if they had to regoal.

7. Share who was in their **Nourishing Network**, and how it helped them navigate the challenge.

8. Pick 3 challenges from the **'Eliminating Challenges'** on the infographic, and share how your hope hero eliminated them.

9. Write the conclusion. What do you want the world to know? What do you wish someone had told you? What is the moral of the story?

If you're inspired, share this hope hero story so we can help activate these skills globally!

#Hope #ShineHope #MyHopeHero

We all experience moments of hopelessness (emotional despair and motivational helplessness). The key is to use the Shine Hope skills to navigate your way from despair to positive feelings, and helplessness to inspired actions. Use the Shine Hope framework to build your muscle.

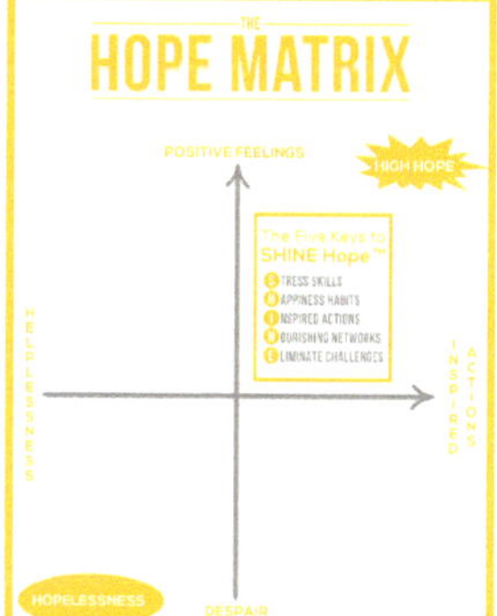

My Hope Hero Template

MY HOPE HERO

Kathryn Goetzke

When Kathryn was 18 years old, a freshman at the University of Iowa, her dad died by suicide. It really changed her life. When she was in her early 20's, she then tried to take her own life, yet didn't tell another soul for 10 years. She knows a lot about hopelessness.

To work on her recovery, she used a lot of Stress Skills. She talks about crying, going to therapy, learning to meditate, deep breathing, and listening to music. She traveled a lot, and took up hiking and exercise. She also took up boxing and spent a lot of time in nature.

Kathryn was diligent about her Happiness Habits. She listened to her favorite band the Killers, went to concerts, focused on her nutrition and sleep, and started exercising. She pursued her passions, started a nonprofit iFred, and did a lot of volunteer work. She got serious about her purpose.

Kathryn also took a lot of Inspired Actions towards her goals. She chunked them down, got a degree and then an MBA. She couldn't talk to her dad anymore, so she found business mentors. Her brothers were always there to support her, and her mom was a source of strength and inspiration.

Kathryn spent a lot of time with her Nourishing Networks. She spent time with people that were kind, compassionate, fun, and helped her heal. She had a therapist and got close to God. She had animals and spent a lot of time with wild horses in Nevada.

She worked to Eliminate Challenges like her rumination and worry. She learned about sensory engagement, and even started a company to teach others. She worked to forgive herself and others. She focused on what she could control, which was her present and future, and did her best to let go of the rest. She put all her failures into teaching others.

Her use of the Shine Hope framework led her on a much healthier path. She has been sober almost 20 years, and had her nonprofit that same amount of time. She is a representative at the United Nations for the World Federation for Mental Health, and has shared her story around the world at places like the World Bank, Harvard, the United Nations, and more. She has created programming to teach hope to kids, published papers, and is now doing workplace programming, has a college, course, and is activating cities. She is on a mission to ensure all know how to hope, one person at a time. She is an inspiration, and someone that truly lives by example practicing all she teaches.

#Hope #ShineHope #MyHopeHero

My Hope Hero Template

MY HOPE HERO

#Hope #ShineHope #MyHopeStory

My Hope Hero Template

City, State, Business, or School Proclamation Template

WHEREAS, research has shown that Hope is teachable, and people with higher levels of hope are more likely to achieve their goals, subsequently improving their well-being. Higher Hope is associated with benefits in all areas of life, such as improved health outcomes (i.e., decreased risk of chronic health conditions and improved treatment adherence), improved academic outcomes (i.e., increased graduation rates and academic performance), improved workplace retention and productivity, reduced anxiety and depression, and improved social connectedness; and

WHEREAS, Hope is a fundamental part of the human experience, reflected in diverse traditions and spiritual beliefs across the world. With approximately 85% of the global population identifying with a religion, hope serves as a universal force that brings people together, fostering resilience, connection, and a shared commitment to a brighter future for all; and

WHEREAS, the United Nations has established a resolution to achieve the 17 Sustainable Development Goals, recognizing that persistent challenges in reaching goals can contribute to feelings of hopelessness, which may impact mental well-being. Equipping individuals with strategies to navigate setbacks and maintain a sense of hope is essential to advancing the 2030 Agenda for Sustainable Development and supporting progress toward the Sustainable Development Goals (SDGs); and

WHEREAS, hopelessness–defined as emotional despair and motivation helplessness – is persistent within 3 in 5 young girls and over 40% of adolescents. It is learned and includes emotional despair and motivational helplessness; and

WHEREAS, hopelessness is a primary symptom of depression and anxiety, and a risk factor for, addiction, violence, self-harm, homelessness, early death, failure to graduate from school, poor health, and underperformance in the workplace. It is also a stronger predictor of suicide than loneliness, as hopelessness maintains loneliness through the prevention of connection and social support; and

WHEREAS, Hope emerges as a powerful unifying force in a time of isolation and conflict. Proactive education and skill-building in cultivating Hope should be recognized as an essential tool enabling individuals to manage challenges, set goals, and take positive actions toward a better future;

Hope Month Mayoral Proclamation Template

NOW, THEREFORE, I, (Your Name), (Your Title) of (City/State/Company/School Name) do hereby declare

April as

MONTH OF HOPE

And courage all individuals, organizations, and communities in (City/State/Company/School Name) to observe this time through educational initiatives, public marketing campaigns, and activities that inspire and strengthen Hope within all.

IN WITNESS WHEREOF, I have set my hand and caused the Seal of the (City/State/Company/School Name), (State), to be affixed this X day of X 2025.

____

(Your Name), (Your Title)
(City/State/Company/School Name)

____

Attest:
(Witness Name), (Witness Title)
(City/State/Company/School Name)

Hope Month Mayoral Proclamation Template

City, State, Business, or School Proclamation Template

WHEREAS, research has shown that Hope is teachable, and people with higher levels of hope are more likely to achieve their goals, subsequently improving their well-being. Higher Hope is associated with benefits in all areas of life, such as improved health outcomes (i.e., decreased risk of chronic health conditions and improved treatment adherence), improved academic outcomes (i.e., increased graduation rates and academic performance), improved workplace retention and productivity, reduced anxiety and depression, and improved social connectedness; and

WHEREAS, Hope is a fundamental part of the human experience, reflected in diverse traditions and spiritual beliefs across the world. With approximately 85% of the global population identifying with a religion, hope serves as a universal force that brings people together, fostering resilience, connection, and a shared commitment to a brighter future for all; and

WHEREAS, the United Nations has established a resolution to achieve the 17 Sustainable Development Goals, recognizing that persistent challenges in reaching goals can contribute to feelings of hopelessness, which may impact mental well-being. Equipping individuals with strategies to navigate setbacks and maintain a sense of hope is essential to advancing the 2030 Agenda for Sustainable Development and supporting progress toward the Sustainable Development Goals (SDGs); and

WHEREAS, hopelessness–defined as emotional despair and motivation helplessness – is persistent within 3 in 5 young girls and over 40% of adolescents. It is learned and includes emotional despair and motivational helplessness; and

WHEREAS, hopelessness is a primary symptom of depression and anxiety, and a risk factor for, addiction, violence, self-harm, homelessness, early death, failure to graduate from school, poor health, and underperformance in the workplace. It is also a stronger predictor of suicide than loneliness, as hopelessness maintains loneliness through the prevention of connection and social support; and

WHEREAS, Hope emerges as a powerful unifying force in a time of isolation and conflict. Proactive education and skill-building in cultivating Hope should be recognized as an essential tool enabling individuals to manage challenges, set goals, and take positive actions toward a better future;

WHEREAS, the United Nations recently passed a resolution declaring July 12th as the International Day of Hope;

International Day of Hope Mayoral Proclamation Template

NOW, THEREFORE, I, (Your Name), (Your Title) of (City/State/Company/School Name) do hereby declare

The 12th day of July

INTERNATIONAL DAY OF HOPE

And courage all individuals, organizations, and communities in (City/State/Company/School Name) to observe this time through educational initiatives, public marketing campaigns, and activities that inspire and strengthen Hope within all.

IN WITNESS WHEREOF, I have set my hand and caused the Seal of the (City/State/Company/School Name), (State), to be affixed this X day of X 2025.

____

(Your Name), (Your Title)
(City/State/Company/School Name)

____

Attest:
(Witness Name), (Witness Title)
(City/State/Company/School Name)

International Day of Hope Mayoral Proclamation Template

Draft Resolution proposed for the General Assembly
International Day of Hope

The General Assembly,

Recalling its previous resolution 66/281 of 12 July 2012, which invites Member States to pursue a more inclusive, equitable and balanced approach to economic growth that promotes sustainable development, poverty eradication, happiness, and the well-being of all peoples,

Recognizing that individuals with higher levels of hope are more likely to achieve their goals[1], subsequently improving their well-being, and that goals are met because these individuals have high agency-related hope thoughts[2] (i.e., belief that they can attain their goals and are successful in life) and pathways-related hope thoughts[2] (i.e., belief that they can overcome barriers and develop alternative solutions to goal a when needed).

Emphasizing that the knowledge of how to proactively management of hopelessness, and learn skills to move towards hope, should be a fundamental human right, that hope is teachable and learnable[3,4], and that teaching hope leads to the motivation to set and pursue goals, take risks, and initiate action[5] which are all skills that are critical to attaining all goals in the Sustainable Development Goals (SDGS),

Recognizing that hopelessness is growing, and predictive of weapon carrying on school property[6], self-harm[6], violence,[6] addiction[7], risky behaviors[8], motor vehicle accidents[9], psychological distress[10], depression[11], and suicide[12], that hopelessness is often a consequence of oppression and discrimination[13] so often higher in vulnerable populations, and that not teaching others how to proactively manage hopelessness

---

[1] Moss, S. A. (2018). *Hope and goal outcomes: The role of goal-setting behaviors* [Master's thesis, Ohio State University]. OhioLINK Electronic Theses and Dissertations Center. http://rave.ohiolink.edu/etdc/view?acc_num=osu1513865199503514

[2] Oettingen, G., & Gollwitzer, P. (2002). Turning hope thoughts into goal-directed behavior. *Psychological Inquiry, 13*(4), 304-307. https://www.jstor.org/stable/1448874

[3] Kirby, K., Sweeney, S., Armour, C., Goetzke, K., Dunne, M., Davidson, M., & Belfer, M. (2021). Developing hopeful minds: Can teaching hope improve well-being and protective factors in children? *Child Care in Practice, 28*(4), 504-521.

[4] Bryce, C., Goetzke, K., O'Brien, V., Espnoza, P., & Tomasulo, D. (2024). Promoting hope: preliminary investigation in a college-level hope curriculum. *Journal of American College Health, 5,* 1-7.

[5] Ojala, M. (2023). Hope and climate-change engagement from a psychological perspective. *Current Opinion in Psychology, 49,* 101514.

[6] Duke, N., Borosky, I., Pettingell, S., & McMorris, B. (2011). Examining youth hopelessness as an independent risk correlate for adolescent delinquency and violence. *Maternal and Child Health Journal, 15*(1), 87-97. doi: 10.1007/s10995-009-0550-6

[7] Jalilian, F., Matin, B., & Ahmadpanah, M. (2014). Substance abuse among college students: Investigating the role of hopelessness. *Life Science Journal, 11*(9), 396-399.

[8] Kelly, D., Rollings, A., & Harmon, J. (2005). Chronic self-destructiveness, hopelessness, and risk-taking in college students. *Psychological Reports, 96*(3). https://doi.org/10.2466/pr0.96.3.620-624

[9] Alavi, S., Mohammadi, M., Souri, H., Kalhori, S., Jannatifard, F., & Sepahbodi, G. (2017). Personality, driving behavior and mental disorders factors as predictors of road traffic accidents based on logistic regression. *Iran Journal of Medical Sciences, 42*(1), 24-31.

[10] Lin, T., Yi, Z., Zhang, S., & Veldhuis, C. (2022). Predictors of psychological distress and resilience in the post-COVID-19 era. *International Journal of Behavioral Medicine, 29,* 506-516.

[11] Rholes, W., Rikind, J., & Neville, B. (2011). The relationship of cognitions and hopelessness to depression and anxiety. *Social Cognition, 3*(1). https://doi.org/10.1521/soco.1985.3.1.36

[12] Wolfe, K., Nakonezny, P., Owen, V., Rial, K., Moorehead, A., Kennard, B., & Emslie, G. (2020). Hopefulessness as a predictor of suicidal ideation in depressed male and female adolescent youth. *Suicide and Life-Threatening Behavior, 49*(1), 253-263. doi: 10.1111/sltb.12428

[13] Mitchell, U., Gutierrez-Kapheim, m., Nguyen, A., & Al-Amin, N. (2020). Hopelessness among middle-age and older Blacks: The negative impact of discrimination and protecting power of social and religious resources. *Innovative Aging, 4*(5), igaa044. doi: 10.1093/geroni/igaa044.

V. 03/2024

UN Resolution Template

impacts our collective ability to implement the 2030 Agenda for Sustainable Development, including the Sustainable Development Goals (SDGs),

Recognizing that hope has relevance to promoting benefits in all areas of life (i.e., health[14], academic[15] and work performance[16], all SDG attainment, relationships[17], etc.), and that there is an importance of their recognition in public policy objectives and related outcomes,

Recognizing the necessity for a more inclusive, equitable, and balanced approach to economic growth that promotes sustainable development, poverty eradication, health, and prosperity of all peoples, and attaining our individual and collective hope,

1. Decides to proclaim May 1st the International Day of Hope;

2. Invites all individuals, Member States, organizations of the United Nations system and other international and regional organizations, as well as civil society, including non-governmental organizations and individuals, to observe the International Day of Hope in an appropriate manner, including through education and public awareness-raising activities;

3. Requests the Secretary-General to bring the present resolution to the attention of all Member States, organizations of the United Nations system, and civil society organizations for appropriate observance. INSERT NUMBER plenary meeting INSERT DATE.

---

[14] Long, K. N., Kim, E. S., Chen, Y., Wilson, M. F., Worthington Jr, E. L., & VanderWeele, T. J. (2020). The role of Hope in subsequent health and well-being for older adults: An outcome-wide longitudinal approach. *Global Epidemiology*, *2*, 100018. https://doi.org/10.1016/j.gloepi.2020.100018

[15] Day, L., Hanson, K., Maltby, J., Proctor, C., & Wood, A. (2010). Hope uniquely predicts objective academic achievement above intelligence, personality, and previous academic achievement. *Journal of Research in Personality*, *44*(4), 550-553. https://doi.org/10.1016/j.jrp.2010.05.009

[16] Yadav, G., & Kumar, S. (2016). Hope: A tool for managing adversities at workplace. *Indian Journal of Health & Wellbeing, 7*(12), 1156.

[17] Stevens, E., Guerrero, M., Green, A., & Jason, L. A. (2018). Relationship of hope, sense of community, and quality of life. *Journal of Community Psychology*, *46*(5), 567-574. https://doi.org/10.1002/jcop.21959

UN Resolution Template

[Insert Date]
[Insert Mayor/Governor/CEO/Principal Name]
[Insert Mayor/Governor/CEO/Principal Office Address]

Dear [Insert Mayor/Governor/CEO/Principal Name],

I'm writing today to ask for your support in making [Insert City/State/Company/School] a Hopeful [City/State/Company/School]. Hopelessness is rising at unprecedented rates. Hopelessness, the feeling of emotional despair and sense of motivational helplessness, is predictive of weapon-carrying on school property, violence, crime, and so much more that negatively impacts our city. Higher hope, on the other hand, is linked to increased community engagement, more productive workforces, better relationships, higher graduation rates, longer lifespans, less crime, lower addiction, and reduced poverty.

We ask you to join the Hopeful Cities Movement by signing a proclamation to make May 1st as the International Day of Hope. The goal of the proclamation is to activate hope in our community by inspiring educators to utilize the free Hopeful Minds curriculums to teach children critical hope skills, encouraging governments, businesses, and schools to participate in the Five Day Global Hope Challenge to teach the Science of Hope, and to encourage all to learn about the Five Keys to SHINE Hope™ and to activate hope in their lives.

iFred, the International Foundation for Research and Education on Depression, has been researching and teaching the Science of Hope worldwide since 2013. They created Hopeful Minds (www.hopefulminds.org), the first free, global program to teach the "how-to" of hope. iFred also created the Hopeful Cities Playbook, a free, sustainable, easily implementable tool that can be used to operationalize the Global Movement for Hope in our city.

The Hopeful Cities Playbook includes methods for activating hope through Science, Education, Government, Art, Awareness, and Workplace. It includes the Adult Snyder Hope Scale, a free tool that measures individual hope, as well as marketing campaigns, free curriculums, mural ideas, gardens, proclamation language, social media kits, and more. The full Hopeful Cities Playbook is available free for download at www.hopefulcities.org, or available for purchase on Amazon.

Everyone is affected by hopelessness at some point in their life. If not taught skills, many react in violent or self-destructive ways, instead of learning how to proactively manage despair and using it to drive change in society. Activating hope within your community will give everyone the tools they need to proactively manage hopelessness and come back to hope, no matter what life brings.

By joining the Hopeful Cities Movement, our [City/State/Company/School] works to not just talk about hope in the abstract, but to educate, inspire, and teach the "how-to" of hope. We must combat the rise of violence, crime, suicide, anxiety, and depression by ensuring that all are taught these critical skills. Our Global Hope Movement starts with you.

Sincerely,

[Insert Your Name]

Letter to Government Leaders

We've identified an International Day of Hope, similar to what they have done for peace, happiness, the environment, and other important issues. As Hope is key to mental health, it is only fitting that the International Day of Hope is in April, the Month of Hope.

We ask all schools, governments, and others to designate July 12th as the International Day of Hope, and create local activities to create awareness.

We use July 12th so that the remainder of the five-day week can be used for Hope activation in schools and workplaces. We encourage you to honor the International Day of Hope in your community with events that spread and activate Hope, such as:

- Planting sunflowers
- Running Public Service Announcements
- Hosting fundraising events
- Unveiling murals
- Teaching Hopeful Minds in classrooms, after school programs, and shelters
- Opening Art Exhibits
- Providing local presentations and events
- Completing workplace challenges

We also have an International Day of Hope online event, showcasing Science, Stories, and Strategies for Hope. It is free and available to anyone around the world. Join the event to listen to the latest Hope Science and hear what others around the world are doing to activate Hope.

You can find out more at www.internationaldayofHope.org.

International Day of Hope

Public Service Announcements

A Public Service Announcement (PSA) is a short 30- to 60-second message that radio and television stations run at no cost. They do this to support the community and serve the public interest. They are a cost-effective way to raise awareness in the community.

Radio stations receive PSAs as audio files, typically recorded as 30- or 60-second messages, that are created by local groups. We created PSAs for our launch in Reno, and encourage you to do the same. You can either create and produce them yourselves, or send the scripts to radio and TV stations and ask them for support in doing so.

We encourage you to use local celebrities and talent to support you in the PSAs, as the more trusted the person delivering the message, the more likely your supporters are to act. Trust is key when developing the Science of Hope.

You can check out the PSAs we did in Reno, and see how we engaged local champions to share the message. If you are interested in making a PSA for your community, simply download the video template provided, and ask your friends, family, local celebrities, and Hope champions to record their message and voice to help you #SpreadHope.

Get your town to take the Five-Day Global Hope Challenge. Get the community measuring their Hope, practicing Hope skills, and teaching Hopeful Minds in schools and afterschool programs. Let everyone know that no matter what anyone faces, they are not alone, and there is help.

Public Service Announcements

Sample Hopeful Cities Public Service Announcements (PSAs)

PSA 1:	PSA 1 (Spanish Version):
Did you know that hope is teachable? That hope is a skill that you can practice? That we can all practice? Did you know that hope predicts your GPA, Sports performance, And your health? Check out the free Five-Day Global Hope Challenge At www.hopefulcities.org 5 days, 5 lessons, 5 actions. And sign up to take the free Global Hope Challenge At Hopefulcities.org	¿Sabías que la esperanza se puede enseñar? ¿Que esperanza es una habilidad que puedes practicar? ¿Que todos podemos practicar? ¿Sabías que la esperanza predice tu GPA, Tu rendimiento deportivo Y tu salud? Visitarte al Desafío de Esperanza Global de 5 días gratis En www.hopefulcities.org 5 días, 5 lecciones, 5 acciones. Y regístrate para participar en el Desafío de Esperanza Global de 5 días gratis En Hopefulcities.org
PSA 2:	**PSA 2 (Spanish Version):**
Did you know Reno is the first-ever Hopeful City in the world! Did you know Hope is a science? Did you know that you can measure hope? Visit HopefulCities.org Get local resources to support you in these challenging times. It's okay to reach out and ask for help At www.hopefulcities.org Join us today.	Sabías ¡Reno es la primera ciudad esperanzada del mundo! ¿Sabías que la esperanza es una ciencia? ¿Sabías que puedes medir la esperanza? Visite HopefulCities.org Obtenga recursos locales para apoyarlo en estos tiempos desafiantes Está bien comunicarte y pedir ayuda En www.hopefulcities.org Únete a nosotros hoy.
PSA 3:	**PSA 3 (Spanish Version):**
Did you know that Hope is teachable? That Hope is a skill that you can practice? That we can all practice? Did you know that Hope predicts your GPA? And your health? Hope is Measurable Learnable Teachable Hopeful Mindsets on the College Campus is a program that teaches you The Hope skills you need to Create, maintain, and grow hope. When you feel lonely, sad, stressed, or overwhelmed Hope can help.	¿Sabías que esperanza es enseñable? ¿Que esperanza es una habilidad que puedes practicar? ¿Que todos podemos practicar? ¿Sabías que esperanza predice tu GPA? ¿Y tu salud? la esperanza es Medible Aprendible Educable Hopeful Mindsets on the College Campus es un programa que te enseña las habilidades de esperanza que necesita para crear, mantener y crecer la esperanza. Cuando te sientes solo, triste, estresado o abrumado la esperanza puede ayudar.

Public Service Announcements

Sample Press Release:

Hopeful Cities

INSERT CITY Commits to Hope, Joining the Hopeful Cities Movement to Ensure the Community Knows the What, Why, and How of Hope

INSERT CITY / STATE / INSERT WIRE / INSERT DATE / The Shine Hope Company (TSHC) proudly announces INSERT CITY joins the Hopeful Cities movement. The Hopeful Cities initiative is a comprehensive program aimed at raising awareness about the elements for creating, maintaining, and growing hope through targeted interventions across various city sectors, including government, the workplace, art, science, education, and healthcare. Hope is measurable, and teachable, and all in INSERT CITY are encouraged to visit www.hopefulcities.org/COUNTRY/STATE/CITY to learn more.

INSERT NAME OF FUNDER OR INFLUENTIAL PERSON IN CITY (i.e.): "Hopelessness is learned and impacts all areas of life", says INSERT NAME OF PERSON. "The physical isolation triggered many complications, including depressive symptoms, anxiety, stress, sleep disorders, and emotional disturbance, and it is easy to fall into hopelessness. The job and home losses are also adding to the dire circumstances of many, and it is critical we use hope as a strategy, and the related skills, to proactively combat the many challenges we face as hope is a protective factor for anxiety and depression, and teachable."

Hopeful Cities aims to operationalize the work of Hope in INSERT CITY through a public health campaign, workplace posters, Hopeful Minds Overview programs for youth, programs for teens, and a Parent's Guide for using hope language at home. Hopeful Cities includes a Hopeful Cities Playbook, a resource with a guide on how to amplify hope science in the following Six Sectors: government, science, healthcare, education, art, and the workplace.Hopeful Cities' Eight Guiding Principles are embedded into every one of our interventions across all six sectors: Take a Whole Community Approach, Bridge the Knowledge-Action Gap, Utilize Solution-Focused Methods, Act Early Everywhere, Empower All Citizens, Amplify a Universal Brand, Use Evidence Informed and Evidence-Based, and Activate the Shine Hope Framework.

INSERT QUOTE FROM MAYOR OR CITY LEADER (i.e.): "I've always thought about hope as a wish, and not understanding that it is something you can learn," INSERT MAYOR'S NAME. "It is exciting to know hope can be taught, and it is something we need to teach all in INSERT CITY. We are proud to be a part of this initiative, and excited to get the work in the hands of those that can help spread the word."

The Hopeful Minds Overview curriculum and Parent's Guide, available at no cost to parents and educators, offers downloadable resources suitable for both classroom and

Press Release Template

remote learning environments. While initially tailored for grades K-6, the curriculum is adaptable for all age groups and was recognized as an innovation by the World Bank. It features Hope Hero stories, including figures like John Krasinski, Magic Johnson, and Selena Gomez, who exemplify the skills of hope. The curriculum is designed for remote learning compliance, aligns with National Health Education Standards, and includes a Parent's Guide to facilitate learning hope language at home. Covering a comprehensive range of topics, the curriculum teaches the significance and potency of hope, mindfulness, emotional self-regulation, gratitude, the brain, nutrition, the concept of success as a journey, handling failure, overcoming rumination and worry, finding purpose, experiencing wonder and awe, navigating change, creating a network for hope, and much more.

Myron Belfer, MD, MPA, an advisor for the Hopeful Minds and, and Professor of Psychiatry in the Department of Psychiatry, Children's Hospital Boston, Harvard Medical School, has shown from a review of research that, "Hope is tangible and teachable, and it is an essential ingredient for a successful life trajectory." He has cited the clinical approach of the Nobel Prize winner Dr. Joseph E. Murray, who helped his severely ill patients not by talking about prognosis but by offering a future orientation embodying hope.

Recognizing hopelessness as the primary predictor of suicide and a key symptom of depression and anxiety, Kathryn Goetzke, MBA, Founder of iFred and CEO and Chief Hope Officer of The Shine Hope Company, transformed theory into action. She pioneered the first-ever free global curriculum explicitly designed to impart hope as a skill. Higher levels of hope correspond to greater emotional and psychological well-being, greater economic security, improved academic performance, less violence, more connection, less loneliness, and enhanced personal relationships.

The Hopeful Cities workplace campaign includes the Five Keys to Shine Hope including; Stress Skills, Happiness Habits, Inspired Actions, Nourishing Networks, and Eliminating Challenges. The free 5-Day Global Hope Challenge includes a daily email, with a video / written lesson that includes one action step to practice daily to start learning about 'how' to hope. There are also links to the Children's and Adult Snyder Hope Scales, so individuals can measure their own journey to hope. The program also includes PSAs, billboards, ads, yard signs, social media kits, to ensure all know what hopelessness is, and be equipped with skills to Shine Hope.

About The Shine Hope Company:

Our mission is to empower all by teaching scientifically informed and evidence-based methods that cultivate hope. Through our educational resources, tools, and activations, we aim to inspire individuals to foster positive feelings and take inspired action, measuring their progress in nurturing and sustaining hope in their lives. Visit www.theshinehopecompany.com to find out more.

Press Release Template

About iFred:

iFred, a 501(c)3 organization, is working to teach hope. iFred has worked to shine a positive light on mental health and eliminate stigma through prevention, research and education and created a shift in society's negative perception of the disease through positive imagery, rebranding, celebrity engagement, cause marketing campaigns, and establishing the sunflower and color yellow as the international symbols for hope. iFred worked with The Mood Factory to do the first nationwide cause marketing campaign for mental health in the US, and created the first ever program to teach hope, based on research it is a teachable skill.

About Hopeful Cities

Hopeful Cities is equipping cities around the world with the tools they need to create, maintain, and grow hope, citywide. This initiative features the Hopeful Cities Playbook, a step-by-step guide to help cities activate the "how-to" of hope in the sectors of government, education, science, workplace, healthcare, and art. The Eight Guiding Principles of Hopeful Cities are integrated into all interventions across these six sectors, driving change and ensuring everyone can proactively move from hopelessness to hope. Visit www.hopefulcities.org to get a copy of the Hopeful Cities Playbook and find out more how you can activate hope in your city.

Media Contacts:

SOURCE: INSERT SOURCE

Press Release Template

Hope Scales

Survey Form

Press Talking Points

1. **Hope Science**
   a. Hope is measurable and teachable. We use Snyder's Children and Adult Hope Scales to measure hope.
   b. Chicago residents can access a range of tools designed to increase hope, that use the Shine Hope Framework of Stress Skills, Happiness Habits, Inspired Actions, Nourishing Networks, and Eliminating Challenges.
   c. Resources include:
      i. The **Children and Adult Hope Scales**, which allow individuals to measure their hope, and the **Shine Hope Infographic**, featuring clickable links to information on the Five Keys to Shine Hope.
      ii. The **Five-Day Global Hope Challenge** introduces the essentials of hope and its measurable impact, while digital posters and brochures are available for workplaces, hospitals, libraries, schools, places of worship, community centers, and other gathering spaces.
      iii. The **My Shine Hope Story** and **My Shine Hope Hero Templates** encourage individuals to share experiences of overcoming challenges and share examples using Shine Hope Skills.
      iv. The **Hopeful Minds Parent's Guide** provides practical tips for practicing skills to create a hope-filled home.
      v. The **Overview and Deep Dive Educator Guides and Workbooks and Hopeful Minds for Teens Program** offer lessons on the "what," "why," and "how" of hope for youth, designed for use by educators, places of worship, after-school programs, police working with youth, community centers, parents, or nonprofits.
      vi. A **Hopeful Mindsets for Veterans Facilitator and Workbook** for use with those that served in the military.
      vii. All programs meet the National Health Education Standards.
2. **Encourage Community Member Involvement**
   a. Encourage individuals to write their My Shine Hope Story as a way to showcase how they can navigate challenges using the Shine framework and provide a model to others who may be facing similar challenges.
   b. Plant sunflower gardens, as sunflowers are the symbol of hope.
   c. Encourage your schools and workplace to integrate hope teaching into the structure to ensure all have the skills to proactively move from hopelessness to hope.
3. **Higher Hope shows:**
   i. Increased resilience against adversity
   ii. Increased productivity in the workplace by 14%, outperforming productivity based on the worker's intelligence, optimism, and self-efficacy
   iii. Increased employee retention
   iv. Increased sleep quality
   v. Increased thriving work environments
   vi. Increased resilience during crises and challenges often found in the workplace
   vii. Decreased the risk of developing chronic conditions, such as cardiovascular disease
   viii. Decreased and protect against mental illness
   ix. Decreased risky behaviors in youth.
   x. Decreased suicide
   xi. And more
4. **Hope and Sustainable Development Goals (SDGs)**
   a. Highlight the correlation between higher hope levels and the ability to set and achieve the SDGs
   b. Those high in hope have the drive and determination to persevere and meet goals while hopelessness can hinder an individual's ability to set and achieve goals.
5. **The Urgency to Address Hopelessness**
   a. Hopelessness is both emotional despair and motivations helplessness (Abramson et al. 1989).
   b. Everyone experiences moments of hopelessness whether they are small (e.g., being stuck in traffic, forgetting to turn in an assignment) or big (i.e., losing a loved one, being fired) and systems can create hopelessness (oppression, discrimination).
   c. Persistent hopelessness is linked to:
      i. Increased suicidal ideation
      ii. Increased risk of mental health disorders, such as depression and anxiety

Press Talking Points

iii. Increased risky behaviors (e.g., substance use, reckless driving, violence, self-harm, and bullying) in teens and young adults
iv. Increased risk of chronic health outcomes
v. Increased absenteeism at work and school
vi. Impaired job engagement and performance, leading to quicker turnover

6. **Hope across the sectors**
   a. **Science:** Call on scientists to share information on hope science, while studying hope-related outcomes across various populations to improve our understanding of hope. Encourage everyone to measure their hope level using the Snyder Hope Scales.
   b. **Art:** Illustrate how sunflower artwork serves as both a beautiful addition to the city and a resource for promoting hope activation.
   c. **Workplace:** Emphasize the substantial economic costs of hopelessness and the ROI on investing in employee well-being.
      i. Research suggests that every $1 invested into employee well-being has an ROI of $4.
   d. **Education:** Showcase how teaching hope to our youth sets them up for success by instilling the skills needed to manage moments of hopelessness.
   e. **Healthcare:** Emphasize how teaching people how to manage hopelessness may protect against depression and chronic health conditions. Depression is the leading cause of disability worldwide, and in total, poor mental health was estimated to cost the world economy approximately $2·5 trillion per year in poor health and reduced productivity in 2010, a cost projected to rise to $6 trillion by 2030.
   f. **Government:** Encourage government officials to declare the first Monday in May as the International Day of Hope to further encourage hope activation in the city.

7. **Hope across the sectors**
   a. **Science:** Call on scientists to share information on hope science, while studying hope-related outcomes across various populations to improve our understanding of hope. Encourage everyone to measure their hope level using the Snyder Hope Scales.
   b. **Art:** Illustrate how sunflower artwork serves as both a beautiful addition to the city and a resource for promoting hope activation.
   c. **Workplace:** Emphasize the substantial economic costs of hopelessness and the ROI on investing in employee well-being.
      i. Research suggests that every $1 invested into employee well-being has an ROI of $4.
   d. **Education:** Showcase how teaching hope to our youth sets them up for success by instilling the skills needed to manage moments of hopelessness.
   e. **Healthcare:** Emphasize how teaching people how to manage hopelessness may protect against depression and chronic health conditions. Depression is the leading cause of disability worldwide, and in total, poor mental health was estimated to cost the world economy approximately $2·5 trillion per year in poor health and reduced productivity in 2010, a cost projected to rise to $6 trillion by 2030.
   f. **Government:** Encourage government officials to declare the first Monday in May as the International Day of Hope to further encourage hope activation in the city.

Press Talking Points

Hopeful Minds Overview Educator Guide

Three, one-hour lessons that introduce the "what," "why," and "how" of Hope targeting 2nd graders, yet easily adaptable for any age.

The Shine Hope Store: theshineHopestore.com/products/paperback-Hopeful-minds-overview-educator-guide
SchoolHealth: www.schoolhealth.com/Hopeful-minds-overview-Hopework-books
Order for your Library or Bookstore: *(See page 123 for instructions)*

Hopeful Minds Overview Hopework Book

Workbook for students designed to supplement the Hopeful Minds Overview Educator Guide lessons. The Hopework book is also included in the Educator Guide itself. Available in English and Spanish version.

The Shine Hope Store:
(English) - theshineHopestore.com/products/paperback-Hopeful-minds-overview-Hopework-book
(Spanish) - theshineHopestore.com/products/Hopeful-minds-overview-Hopework-book-spanish-edition-print
SchoolHealth: www.schoolhealth.com/Hopeful-minds-overview-Hopework-books
Order for your Library or Bookstore: *(See page 123 for instructions)*

Hopeful Minds Parent's Guide

Provides a broad overview of the concepts discussed in the Hopeful Minds curriculums, and provides parents with easy ways to implement the Five Keys to Shine Hope (Stress Skills, Happy Habits, Inspired Actions, Nourishing Networks, and Eliminating Challenges) and Hopeful language at home.

The Shine Hope Store: theshineHopestore.com/products/paperback-Hopeful-minds-parents-guide
SchoolHealth: www.schoolhealth.com/Hopeful-minds-parent-s-e-guide
Order for your Library or Bookstore: *(See page 123 for instructions)*

Hopeful Minds Curriculums

Hopeful Minds Deep Dive Educator Guide

16, 45-minute lessons and delve into the skills needed to create, maintain, and grow a Hopeful mindset, targeting 5th graders yet easily adaptable for any age.

The Shine Hope Store: theshineHopestore.com/products/paperback-Hopeful-minds-deep-dive-educator-guide
SchoolHealth: www.schoolhealth.com/Hopeful-minds-deep-dive-Hopework-e-books
Order for your Library or Bookstore: *(See page 123 for instructions)*

Hopeful Minds Deep Dive Hopework Book

Workbook for students designed to supplement the Hopeful Minds Deep Dive Educator Guide lessons. The Hopework book is also included in the Educator Guide itself. Available in English and Spanish version.

The Shine Hope Store:
(English) - theshineHopestore.com/products/paperback-Hopeful-minds-deep-dive-Hopework-book
(Spanish) - theshineHopestore.com/products/Hopeful-minds-deep-dive-Hopework-book-spanish-edition-print
SchoolHealth: www.schoolhealth.com/Hopeful-minds-deep-dive-Hopework-e-books
Order for your Library or Bookstore: *(See page 123 for instructions)*

Hopeful Minds Teen Hopeguide

Tested and approved by teens, the 'Hopeful Minds for Teens' program is a 12-module workbook that introduces the Five Keys to Shine Hope™: Stress Skills, Happiness Habits, Inspired Actions, Nourishing Networks, and Eliminating Challenges. This comprehensive approach empowers teens to navigate challenges, embrace positivity, and empower resilience.

The Shine Hope Store: theshineHopestore.com/products/Hopeful-minds-teen-Hopeguide-print
SchoolHealth: www.schoolhealth.com/Hopeful-minds-teen-Hope-e-guide
Order for your Library or Bookstore: *(See page 123 for instructions)*

Hopeful Minds Curriculums

IngramSpark is a print-on-demand and distribution service for independent authors and publishers, allowing them to publish and distribute their books to various bookstores, retailers, and libraries. If you are a bookstore, retailer, or library looking to order books from IngramSpark, you can do so by following these general steps:

Create an Account

1. Visit the Ingram Content website (www.ingramcontent.com).
2. Create an account specific to your business type (retailer, library, etc.). You may need to provide business-related information during registration.

Ingram Content may require some verification to ensure you are a legitimate business or organization.

Order the Books to Sell on your Retail / Book store

Our books are made available on iPage, Ingram's easy-to-use online search, order, and account management platform for titles.

1. Once your account is set up and verified, go to ipage.ingramcontent.com/ipage/auth/login
2. Use the search and browse functions on the website to find the books you want to order. (Refer to the ISBN of our books below)

Checkout

1. When you find the books you want, add them to your cart just like you would with any online retailer.
2. Proceed to the checkout page to review your order.
3. Provide payment information and complete the purchase.
4. Select your preferred shipping method and enter the shipping address where you want the books to be delivered.
5. Pay for your order, and you will receive an order confirmation.
6. Wait for the books to be shipped to your designated address.

Upon receiving the books, manage your inventory accordingly, whether for resale in your bookstore or for circulation in your library.

It's important to note that Ingram Content offers a vast catalog of books from various independent authors and publishers. You can order books in various formats, such as hardcover, paperback, and e-books. Additionally, you can set up ongoing relationships with Ingram Content for continuous ordering and stocking of books as needed.

Keep in mind that specific ordering processes and requirements may vary, so it's a good idea to contact Ingram Content's customer support or refer to their official website for the most up-to-date information and any specific guidelines for your type of business or organization.

How to Order for your Library or Bookstore

DID YOU KNOW? HOPE IS TEACHABLE.

WHY HOPE?

Covid-19 and inequality across the world have created unprecedented stress on parents, teachers, and children. Anxiety and depression can begin to appear by age 7 and will continue to develop through middle school and high school. A recent study found that over 50% of girls in the United States experience hopelessness- a known predictor of anxiety and depression, and the best predictor of suicide.

Hope is the antithesis of hopelessness, and a known protective factor against anxiety, depression, addiction, and suicide. Higher hope is associated with improved athletic abilities, academic achievements, productivity, social connection, and health. By choosing this curriculum, you are choosing hope and taking the first step towards teaching children critical skills that will have lasting, positive impacts on their futures.

WHY HOPEFUL MINDS?

- Hopeful Minds offers global program aimed at teaching hope as a skill to youth around the world.
- The Hopeful Minds resources have been downloaded over 5,000 times by educators around the world. The Hopeful Minds resources include a three lesson Hopeful Minds Overview, a 16 lesson Hopeful Minds Deep Dive, a Hopeful Minds Parent's Guide, a "Resources for Stress, Anxiety, and Depression" booklet, and a "Where to Find Support" booklet.
- Named an Innovation by the World Bank, Hopeful Minds has been featured at the BBC, the United Nations World Federation for Mental Health, the Mental Health Community Associations Conference, the Kennedy Forum, and more.
- All Hopeful Minds curriculums are remote learning adaptive, reinforce the eight National Health Education Standards set forth by the Centers for Disease Control and Prevention (CDC), meet Social Emotional Learning (SEL) guidelines, are Adverse Childhood Experiences (ACES)-informed, and have bene proven to provide effective anti-bullying strategies.
- In addition to interactive lessons that explore the tools needed to activate hope, the curriculums also include Hope Hero spotlights, hope stories, background information for educators, supplemental resources, classroom visuals, and a Hopework Book for students.

Hear what educators are saying:

"We love the materials. We used this program all last year and plan to use it again." - Julie

"Keep finding ways to teach hope! I'm sharing this resource with everyone I know." - Turyn

"Great stuff! Easy to navigate and digest." - Nichole

Learn more at www.hopefulminds.org

 @ifredorg  @ifredorg  @ifredorg

Hopeful Minds Flyer

Hopelessness is emotional despair...
and motivational helplessness, and something that everyone feels at one time or another. When left unchecked, hopelessness can lead to destructive behaviors and negative life outcomes that cause harm to ourselves and others.
This is where Hope can help.
Hope can be created, maintained, and grown using the Five Keys to Shine Hope™.
tress Skills
appiness Habits
nspired Actions
ourishing Networks
liminating Challenges
the shine hope™ company
RESOURCES FOR HOPE
In addition to your Hope Network, there are also resources for hope that you can use when you are experiencing hopelessness. If you need immediate assistance, you can use any of the resources below to talk to someone in your local area.
FIND WHERE TO GET SUPPORT AT:
www.ifred.org
GET PEER TO PEER TRAINING AT:
www.youthera.org
MEASURE YOUR HOPE:
Scan the QR Code
©2023, The Shine Hope Company
SHINE HOPE
A How-To for HOPE in Trying Times

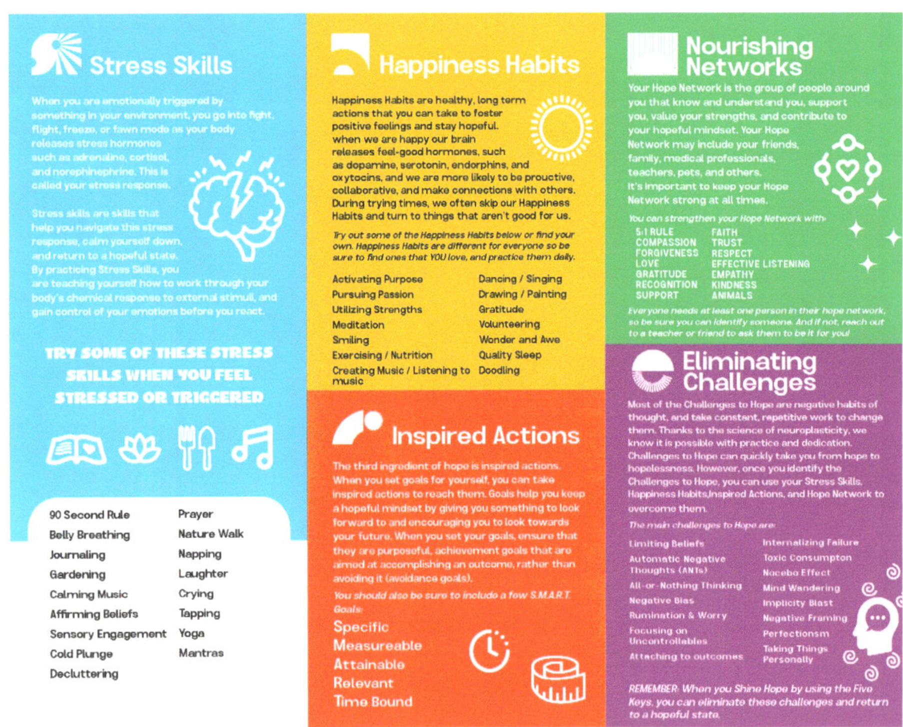

Stress Skills
When you are emotionally triggered by something in your environment, you go into fight, flight, freeze, or fawn mode as your body releases stress hormones such as adrenaline, cortisol, and norephinephrine. This is called your stress response.
Stress skills are skills that help you navigate this stress response, calm yourself down, and return to a hopeful state. By practicing Stress Skills, you are teaching yourself how to work through your body's chemical response to external stimuli, and gain control of your emotions before you react.
TRY SOME OF THESE STRESS SKILLS WHEN YOU FEEL STRESSED OR TRIGGERED
90 Second Rule
Belly Breathing
Journaling
Gardening
Calming Music
Affirming Beliefs
Sensory Engagement
Cold Plunge
Decluttering
Prayer
Nature Walk
Napping
Laughter
Crying
Tapping
Yoga
Mantras
Happiness Habits
Happiness Habits are healthy, long term actions that you can take to foster positive feelings and stay hopeful. when we are happy our brain releases feel-good hormones, such as dopamine, serotonin, endorphins, and oxytocins, and we are more likely to be prouctive, collaborative, and make connections with others. During trying times, we often skip our Happiness Habits and turn to things that aren't good for us.
Try out some of the Happiness Habits below or find your own. Happiness Habits are different for everyone so be sure to find ones that YOU love, and practice them daily.
Activating Purpose
Pursuing Passion
Utilizing Strengths
Meditation
Smiling
Exercising / Nutrition
Creating Music / Listening to music
Dancing / Singing
Drawing / Painting
Gratitude
Volunteering
Wonder and Awe
Quality Sleep
Doodling
Inspired Actions
The third ingredient of hope is inspired actions. When you set goals for yourself, you can take inspired actions to reach them. Goals help you keep a hopeful mindset by giving you something to look forward to and encouraging you to look towards your future. When you set your goals, ensure that they are purposeful, achievement goals that are aimed at accomplishing an outcome, rather than avoiding it (avoidance goals).
You should also be sure to include a few S.M.A.R.T. Goals:
Specific
Measureable
Attainable
Relevant
Time Bound
Nourishing Networks
Your Hope Network is the group of people around you that know and understand you, support you, value your strengths, and contribute to your hopeful mindset. Your Hope Network may include your friends, family, medical professionals, teachers, pets, and others. It's important to keep your Hope Network strong at all times.
You can strengthen your Hope Network with:
5:1 RULE
COMPASSION
FORGIVENESS
LOVE
GRATITUDE
RECOGNITION
SUPPORT
FAITH
TRUST
RESPECT
EFFECTIVE LISTENING
EMPATHY
KINDNESS
ANIMALS
Everyone needs at least one person in their hope network, so be sure you can identify someone. And if not, reach out to a teacher or friend to ask them to be it for you!
Eliminating Challenges
Most of the Challenges to Hope are negative habits of thought, and take constant, repetitive work to change them. Thanks to the science of neuroplasticity, we know it is possible with practice and dedication. Challenges to Hope can quickly take you from hope to hopelessness. However, once you identify the Challenges to Hope, you can use your Stress Skills, Happiness Habits,Inspired Actions, and Hope Network to overcome them.
The main challenges to Hope are:
Limiting Beliefs
Automatic Negative Thoughts (ANTs)
All-or-Nothing Thinking
Negative Bias
Rumination & Worry
Focusing on Uncontrollables
Attaching to outcomes
Internalizing Failure
Toxic Consumpton
Nocebo Effect
Mind Wandering
Implicity Blast
Negative Framing
Perfectionsm
Taking Things Personally
REMEMBER: When you Shine Hope by using the Five Keys, you can eliminate these challenges and return to a hopeful state.

Teen Hope Brochure

Hopeful Mindsets on the College Campus is a 10-module video course from The Shine Hope Company that equips students with crucial Hope skills through expert insights and real-life stories. The course features experts from Harvard, Stanford, and Columbia, with insights from recent college graduates that offer real-life practical strategies and stories from their experiences with homelessness, mental health diagnoses, death, violence, and everyday challenges at school.

The video course is available for individual purchase at www.Hopecourses.com. For bulk license, e-mail us at activate@theshinehopecompany.com.

Hopeful Mindsets on the College Campus 10-Module Video Course

Hopeful Mindsets®
on the College Campus

DID YOU KNOW? HOPE PREDICTS RETENTION (AND IS TEACHABLE).

WHY HOPEFUL MINDSETS?

- ***Hope uniquely predicts college retention, a key indicator of school ratings and funding support.*** Research has shown that among first-semester college students, hope predicts second-semester enrollment above and beyond high school academic performance. Hope also predicts objective academic achievement above intelligence, personality, or previous academic achievement.
- Hopeful Mindsets teaches students and educators how to proactively manage stress, channel emotions for good, create goals, overcome obstacles, and create a mindset for hope, so no matter what life brings, there is always a way from hopelessness to hope.
- Hopeful Mindsets on the College campus consists of 10 online video and text lessons, guided coursebook reflections, lesson quizzes, a college marketing campaign, and a CANVAS page outline with suggested campus-specific resources. The course combines interviews with hope experts, including experts from Harvard, Stanford, Columbia, and more, stories from college graduates, and Hope Science to introduce critical hope skills.
- Initial research conducted at Arizona State University said approximately 70% of students who took the Hopeful Mindset College course were very or extremely engaged in the course and 50% stated that they were more engaged in the Hopeful Mindset course than any of their other classes. Initial research suggests for hope, the control group showed a decline in hope levels from pre-test to post test and the HM students showed an increase in hope scores.

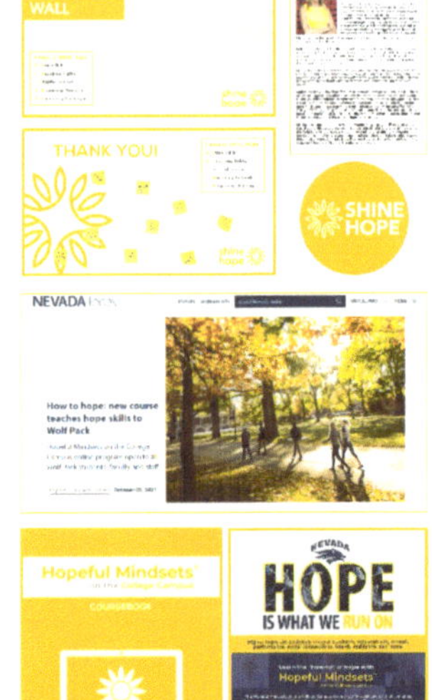

HOPE HERO WALL
THANK YOU!
SHINE HOPE
NEVADA
How to hope: new course teaches hope skills to Wolf Pack
Hopeful Mindsets
HOPE
IS WHAT WE RUN ON

WHY HOPE?

Hope is a known protective factor against anxiety, depression, addiction, and suicide. Hope impacts all outcomes in life, including academic outcomes, athletic performance, health, and resilience. Hope uniquely predicts if a student will return to campus the following year.

WHY NOW?

Hopelessness is growing at unprecedented rates among youth and adults around the world and is the leading cause of suicide and primary symptom of depression. The JED Foundation found that 63% of college students say their emotional health is worse now than before the COVID-19 pandemic. A high proportion of college students are dealing with anxiety (82%), social isolation and loneliness (68%), depression (63%), and difficulty coping with stress in a healthy way (60%). Our ability to effectively manage and adapt to these times determines our success as a society. Join us today by bringing Hopeful Mindsets to your school and community.

Learn more about Hopeful Mindsets by visiting www.hopefulmindsets.com

Hopeful Mindsets on the College Campus Flyer

DO YOU WANT TO BE A

HOPEFUL HOG?

Hope is Measurable and Teachable!

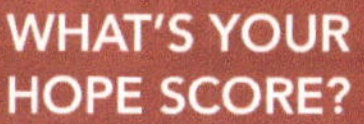

- Hope is a protective factor for violence, drug/alcohol abuse, depression, anxiety, and suicide.
- Higher hope can positively impact academic achievement, athletic performance, social connections, health, resilience, and more.
- This course reveals facts on hope science from experts and personal experiences from graduate students.

Learn the "how-to" of hope by taking

Hopeful Mindsets®

on the College Campus

at no cost!

Use coupon code:
hopefulhogs

**Submit your Shine Hope story to the link at the end of the course*

If you have questions reach out to Avery Peel at **avery@theshinehopecompany.com**

IN CRISIS?
Text HOME to 741741
for immediate assistance.

Brought to you by:

Hopeful Mindsets on the College Campus Flyer (co-branded with Universities)

The Hopeful Mindsets Overview is a 90-minute video course that introduces Hope and the Five Keys to Shine Hope framework to help you create, maintain, and grow Hope in your life.

This course is taught by Kathryn Goetzke, based on her knowledge of mental health and Hope, and her work to date. It compiles knowledge from leading experts on Hope, Mindset, Mental Health, Stress, Positive Psychology, Business, Communications, and more.

Consider using our General Overview course with patients, for peer-to-peer programs, or for staff.

The video course is available for individual purchase at www.Hopecourses.com. For bulk license, e-mail us at activate@theshinehopecompany.com.

Hopeful Mindsets®
General Overview
HOPE MATRIX
Hopeful Mindsets
Kathryn Goetzke
A 90-minute video course that introduces hope and the Five Keys to SHINE Hope™ framework to help you create, maintain, and grow hope in your life.

Hopeful Mindsets General Overview Video Course

The Hopeful Mindsets Overview Workplace Video Course offers an understanding of Hope and its practical application in the workplace. The annual license grants access to the 90-minute video course for employees, providing them with the tools and knowledge to cultivate Hopeful mindsets and promote a positive work environment. We will also provide access to digital Shine Hope Posters that can be used to reinforce the Shine framework throughout the workplace.

The video course is available for individual purchase at www.Hopecourses.com. For bulk license, e-mail us at activate@theshinehopecompany.com.

Hopeful Mindsets®
in the Workplace Overview
Kathryn Goetzke
HOPE MATRIX
STOP BREATHE RELAX
A 90-minute video course that introduces hope and the Five Keys to SHINE Hope™ framework to help you create, maintain, and grow hope in the workplace.

Hopeful Mindsets Overview Workplace Video Course

Take the Challenge.

5 Days. 5 Actions. 5 Lessons.

The Five-Day Global Hope Challenge highlights the critical hope skills encompassed in the Five Keys to **Shine Hope™**. Encourage your community to take the free Five Day Global Hope Challenge and start learning about the science and strategies for hope.

To find free local resources and start activating hope in your life and community, visit www.hopefulcities.org. No matter what life brings, there is always a way from hopelessness to hope.

www.hopefulcities.org
#HopefulCities

Five-Day Hope Challenge Poster

Scan to download the clickable version of this infographic

S TRESS SKILLS

- 90 second pause
- Belly breathing
- Journaling
- Gardening
- Calming music
- Affirming beliefs
- Sensory engagement
- Cold plunge
- Decluttering
- Prayer
- Nature walk
- Napping
- Laughter
- Crying
- Tapping
- Yoga
- Mantras

- Activating purpose
- Pursuing passion
- Utilizing strengths
- Meditation
- Smiling
- Exercising / Nutrition
- Creating / listening to music
- Dancing / Singing
- Drawing / Painting
- Gratitude
- Volunteering
- Wonder / Awe
- Quality sleep
- Doodling

I NSPIRED ACTIONS

- WOOP process
- SMART goals
- Stretch goals
- Achievement goals
- Intrinsic goals
- Mastery goals
- Micro goals / Stepping
- Habit Stacking
- Visualization
- Overcoming obstacles
- Regoaling
- Write down goals / check in

- 5:1 Rule
- Compassion
- Forgiveness
- Love
- Gratitude
- Recognition
- Support
- Faith
- Trust
- Respect
- Effective Listening
- Empathy
- Kindness
- Animals

- Limiting beliefs
- Automatic Negative Thoughts (ANTs)
- All-or-nothing thinking
- Negative bias
- Rumination & Worry
- Focusing on Uncontrollables
- Attaching to outcomes
- Internalizing failure
- Toxic Consumption
- Nocebo Effect
- Mind Wandering
- Implicit Bias
- Negative Framing
- Perfectionism
- Taking things personally

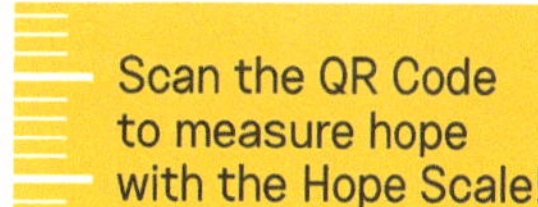

Scan the QR Code
to measure hope
with the Hope Scale!

Shine Hope Posters

Stress Skills are actions that help you navigate your stress response and work through your body's chemical response to external stimuli. By practicing them, you are teaching yourself how to proactively manage the emotional despair found in hopelessness and move towards positive feelings where you activate hope.

The Stress Response

This is when you are emotionally triggered by something in your environment, and you go into fight, flight, freeze, or fawn mode as your body releases stress hormones, such as cortisol, adrenaline, and norepinephrine. You are in your downstairs brain, and can't reach your upstairs brain; the upstairs brain is the place where you make good decisions for moving towards all you hope for in life.

- 90 second pause
- Belly breathing
- Journaling
- Gardening
- Calming music
- Affirming beliefs
- Sensory engagement
- Cold plunge
- Decluttering
- Prayer
- Nature walk
- Napping
- Laughter
- Crying
- Tapping
- Yoga
- Mantras

Shine Hope Posters

Happiness Habits are healthy, long-term actions that cause your brain to release happiness hormones including endorphins, dopamine, serotonin, and oxytocin. Happiness Habits help you stay in your upstairs brain, where you access the problem-solving skills, collaboration, and passion critical for hope.

Positive Feelings

Positive feelings, the first ingredient of hope, are feelings that are located in your upstairs brain like wonder, joy, and peace that make it easier to overcome obstacles that get in the way of hope. You proactively manage the emotional despair of hopelessness using Stress Skills and use your Happiness Habits to stay in your upstairs brain, where you then energetically move towards your goals in life.

- Activating purpose
- Pursuing passion
- Utilizing strengths
- Meditation
- Smiling
- Exercising / Nutrition
- Creating / listening to music
- Dancing / Singing
- Drawing / Painting
- Gratitude
- Volunteering
- Wonder/Awe
- Quality sleep
- Doodling

Shine Hope Posters

Inspired Actions, the second ingredient of hope, are the deliberate steps you take toward your goals in life. Inspired Actions help you to move away from the motivational helplessness, the second ingredient of hopelessness, and toward what you are hopeful for in life.

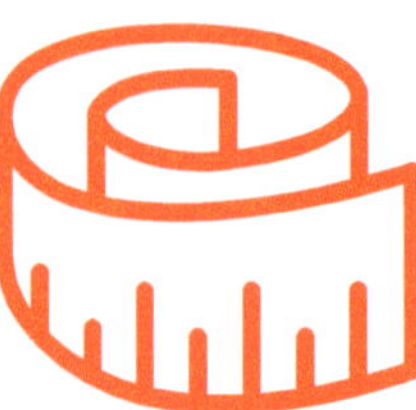

Types of Goals:

WOOP	SMART
Achievement	Stretch
Intrinsic	Micro-Goals

Pathways, Agency, and Regoaling

Obstacles are inevitable, and sometimes you can't reach the goal as you intended. It is important to embrace obstacles to goals, learn to pivot or reevaluate, be flexible and adaptable, and never be afraid to ask for help.

If a goal seems too big, use the stepping process or create micro-goals to chunk it down into smaller goals. Think of one thing you can do in the next 20 minutes. And know when you need to re-goal.

Shine Hope Posters

Your Nourishing Networks, also known as your Hope Networks, are the people in your life that provide you with support, help you stay on track, encourage you to succeed, and who you do the same for in return. You are up to 95% more likely to achieve a goal if you write it down, and check in with someone regularly. So Nourishing Networks are critical support systems for moving you towards what you hope for in life.

Your Hope Networks should include:

People who know and understand you.

People who value your strengths.

People who activate the SHINE framework.

People whom you trust and can confide in.

People who are available to support you.

People you are willing to do the above for as well.

Enhancing Your Hope Networks

Enhance your Hope Networks using the 5:1 rule, vulnerability, praise, recognition, kindness, gratitude, empathy, compassion, collaboration, and strong communication, and be sure to have different networks for different areas of life.

Don't forget to include doctors, therapists, and/or other medical professionals in your Hope Networks.

Shine Hope Posters

Challenges to Hope are negative habits of thought that quickly take you to hopelessness, that emotional despair and sense of helplessness. The thought patterns are often unconscious habits, so becoming aware of these patterns is critical. Once we know what they are and recognize them, it is important to counteract them so that we don't let them keep us from all we hope for in life.

Eliminating Challenges

Most of the Challenges to Hope take constant, repetitive actions to change and overcome. Thanks to the science of neuroplasticity, we know it is possible with practice and dedication. The key is to learn to identify what specific challenges happen most frequently and then proactively find ways to manage those challenges.

Limiting beliefs	Focusing on Uncontrollables	Mind Wandering
Automatic Negative Thoughts (ANTs)	Attaching to outcomes	Implicit Bias
All-or-nothing thinking	Internalizing failure	Negative Framing
Negative bias	Toxic Consumption	Perfectionism
Rumination & Worry	Nocebo Effect	Taking things personally

Shine Hope Posters

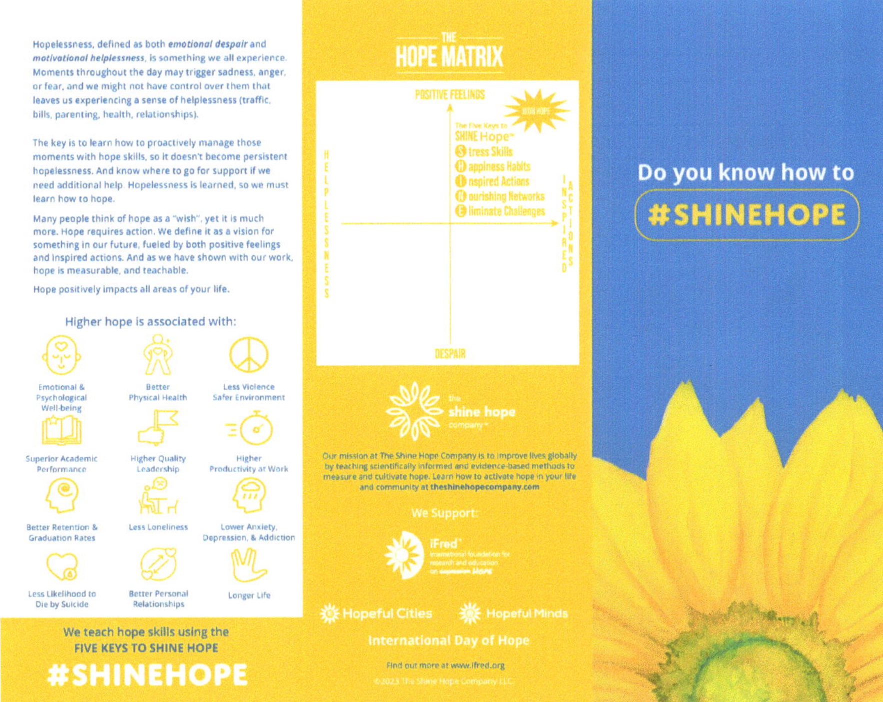

Hopelessness, defined as both emotional despair and motivational helplessness, is something we all experience. Moments throughout the day may trigger sadness, anger, or fear, and we might not have control over them that leaves us experiencing a sense of helplessness (traffic, bills, parenting, health, relationships).
The key is to learn how to proactively manage those moments with hope skills, so it doesn't become persistent hopelessness. And know where to go for support if we need additional help. Hopelessness is learned, so we must learn how to hope.
Many people think of hope as a "wish", yet it is much more. Hope requires action. We define it as a vision for something in our future, fueled by both positive feelings and inspired actions. And as we have shown with our work, hope is measurable, and teachable.
Hope positively impacts all areas of your life.
Higher hope is associated with:
Emotional & Psychological Well-being
Better Physical Health
Less Violence Safer Environment
Superior Academic Performance
Higher Quality Leadership
Higher Productivity at Work
Better Retention & Graduation Rates
Less Loneliness
Lower Anxiety, Depression, & Addiction
Less Likelihood to Die by Suicide
Better Personal Relationships
Longer Life
We teach hope skills using the
FIVE KEYS TO SHINE HOPE
#SHINEHOPE
THE HOPE MATRIX
POSITIVE FEELINGS
HIGH HOPE
The Five Keys to SHINE Hope™
Stress Skills
Happiness Habits
Inspired Actions
Nourishing Networks
Eliminate Challenges
HELPLESSNESS
INSPIRED ACTIONS
DESPAIR
the shine hope company™
Our mission at The Shine Hope Company is to improve lives globally by teaching scientifically informed and evidence-based methods to measure and cultivate hope. Learn how to activate hope in your life and community at theshinehopecompany.com
We Support:
iFred
Hopeful Cities
Hopeful Minds
International Day of Hope
Find out more at www.ifred.org
©2023 The Shine Hope Company LLC
Do you know how to
#SHINEHOPE

The Five Keys to Shine Hope™ highlight skills that are critical for maintaining a sense of hope, no matter what challenges are ahead. Learn more about the Five Keys using the SHINE mnemonic below:
Stress Skills
When you are emotionally triggered by something in your environment, your body has a physiological response in which it releases stress hormones, such as adrenaline, cortisol, and norepinephrine. By practicing Stress Skills, you are teaching yourself how to work through your body's chemical response to external stimuli, and gain control of your emotions before you react. It takes a full 90 seconds for your body to return to equilibrium after your last point of trigger, so when in doubt, take a breath.
Try some of these Stress Skills when you feel stressed or triggered!
• 90-second pause • Belly breathing • Journaling • Gardening • Calming music • Affirming beliefs • Sensory engagement • Cold plunge
• Decluttering • Prayer • Nature Walk • Napping • Laughter • Crying • Tapping
• Yoga • Mantras
Happiness Habits
Happiness Habits are healthy, long term actions that you can take to foster positive feelings and stay hopeful. When you practice Happiness Habits, your body releases serotonin, dopamine, endorphins, and oxytocin, the positive hormones that help counteract stress.
Happiness Habits are different for everyone, so be sure to find ones that YOU love, and practice them daily:
• Activating purpose • Pursuing passion • Utilizing strengths • Meditation • Smiling • Exercising / Nutrition • Creating music / Listening to music
• Dancing / Singing • Drawing / Painting • Gratitude • Volunteering • Wonder / Awe • Quality sleep • Doodling
Inspired Actions
Goals help you stay hopeful by giving you something to look forward to and encouraging you to work towards your future. When you set your goals, you can set a variety of goals yet ensure they are purposeful.
SMART
MEASURABLE
ATTAINABLE
RELEVANT
TIME-BOUND
When you set your goals, set both SMART and stretch goals, and ensure they are purposeful
Nourishing Networks
Your Nourishing Hope Network is the group of people around you that know and understand you, support you, value your strengths, and contribute to your hopeful mindset and includes friends, family, medical professionals, teachers, pets, and others. We can create a network at any age, so even if you can't identify someone now, make a SMART goal as having at least one person to talk is critical for hope. You can strengthen your Hope Network with:
• 5:1 rule • Compassion • Forgiveness • Love • Gratitude
• Recognition • Support • Faith • Trust • Respect
• Effectiv Listening • Empathy • Kindness • Animals
Eliminating Challenges
Challenges to Hope are negative thought patterns that get in the way of our ability to hope. Thanks to neuroplasticity, we can change our brain and stop these patterns. When you encounter challenges to hope, remember to use your Stress Skills, Happiness Habits, Inspired Actions, and Nourishing Networks to eliminate the challenges. The main challenges to hope are:
• Limiting beliefs • Automatic negative thoughts (ANTs) • All-or-nothing thinking • Negative bias • Rumination & Worry • Focusing on uncontrollables • Attaching to outcomes
• Internalizing failure • Toxic Consumption • Nocebo Effect • Mind Wandering • Implicity Bias • Negative Framing • Perfectionism • Taking things personally
And remember, no matter what life brings, Keep Shining! #ShineHope
RESOURCES TO SHINE HOPE
*Scan the QR code to view the full details of the resources below.
• Take the Five Day Global Hope Challenge.
• Listen to the Goal Meditation Audio, a guided meditation to help you think through what you are hopeful in life, and help you create strategies for getting there.
• Take the 90-minute Hopeful Mindsets® General Overview and Workplace course.
• Ask your campus to implement Hopeful Mindsets on the College Campus.
• Fill up your space with Workplace posters, to remember the Shine Hope framework and practice with colleagues.
• Teach Hopeful Minds® in your schools.
• Ask your mayor to become a Hopeful City.
These resources are all available on our website at theshinehopecompany.com
If you or someone you know are having challenges with hope, here are some additional resources:
• Check out our full list of individual support: www.ifred.org/individual-support
• Join our online support group for anxiety and depression at: www.ifred.org
• See if your company offers an Employee Assistance Program (EAP) to utilize benefits like free and confidential clinical counseling services, or gym memberships, coaches, or nutritional counseling.
• Reach out to someone in your Hope Network (or learn how to create one through our Hopeful Mindsets course)
• Contact a healthcare provider (i.e., general practitioner, psychologist, or therapist) who is trained to evaluate, diagnose, and treat mental health challenges.
For urgent support, call 988 or text HOME to 741741 to connect with a crisis counselor at the Crisis Text line.

Shine Hope Brochure

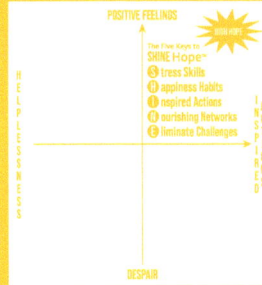

Hopelessness is both emotional despair (sad, anger, fear) **and motivational helplessness** (powerless). **We all experience moments of them, every day.**

Shine skills navigate you back to HOPE.

Scan the code to measure hope with the Hope Scale!

A HOW-TO FOR HOPE IN TRYING TIMES

S TRESS SKILLS

- 90 second pause
- Belly breathing
- Journaling
- Gardening
- Calming music
- Affirming beliefs
- Sensory engagement
- Cold plunge
- Decluttering
- Prayer
- Nature walk
- Napping
- Laughter
- Crying
- Tapping
- Yoga
- Mantras

H APPINESS HABITS

- Activating purpose
- Pursuing passion
- Utilizing strengths
- Meditation
- Smiling
- Exercising / Nutrition
- Creating / listening to music
- Dancing / Singing
- Drawing / Painting
- Gratitude
- Volunteering
- Wonder/Awe
- Quality sleep
- Doodling

I NSPIRED ACTIONS

- WOOP process
- SMART goals
- Stretch goals
- Achievement goals
- Intrinsic goals
- Mastery goals
- Micro goals / Stepping
- Habit stacking
- Visualization
- Overcoming obstacles
- Regoaling
- Write down goals / check in

N OURISHING NETWORKS

- 5:1 Rule
- Compassion
- Forgiveness
- Love
- Gratitude
- Recognition
- Support
- Faith
- Trust
- Respect
- Effective listening
- Empathy
- Kindness
- Animals

E LIMINATING CHALLENGES

- Limiting beliefs
- Automatic Negative Thoughts (ANTs)
- All-or-nothing thinking
- Negative bias
- Rumination & Worry
- Focusing on uncontrollables
- Attaching to outcomes
- Internalizing failure
- Toxic consumption
- Nocebo effect
- Mind wandering
- Implicity bias
- Negative framing
- Perfectionism
- Taking things personally

Hopelessness is both emotional despair (sad, anger, fear) **and motivational helplessness** (powerless). **We all experience moments of them, every day.**

Shine skills navigate you back to HOPE.

A HOW-TO FOR HOPE IN TRYING TIMES

S TRESS SKILLS
90 second pause
Belly breathing
Journaling
Gardening
Calming music
Affirming beliefs
Sensory engagement
Cold plunge
Decluttering
Prayer
Nature walk
Napping
Laughter
Crying
Tapping
Yoga
Mantras

H APPINESS HABITS
Activating purpose
Pursuing passion
Utilizing strengths
Meditation
Smiling
Exercising / Nutrition
Creating / listening to music
Dancing / Singing
Drawing / Painting
Gratitude
Volunteering
Wonder/Awe
Quality sleep
Doodling

I NSPIRED ACTIONS
WOOP process
SMART goals
Stretch goals
Achievement goals
Intrinsic goals
Mastery goals
Micro goals / Stepping
Habit stacking
Visualization
Overcoming obstacles
Regoaling
Write down goals / check in

N OURISHING NETWORKS
5:1 Rule
Compassion
Forgiveness
Love
Gratitude
Recognition
Support
Faith
Trust
Respect
Effective listening
Empathy
Kindness
Animals

E LIMINATING CHALLENGES
Limiting beliefs
Automatic Negative Thoughts (ANTs)
All-or-nothing thinking
Negative bias
Rumination & Worry
Focusing on uncontrollables
Attaching to outcomes
Internalizing failure
Toxic consumption
Nocebo effect
Mind wandering
Implicity bias
Negative framing
Perfectionism
Taking things personally

©2023 The Shine Hope Company LLC

Shine Hope Infographic

Cause marketing involves a collaboration between a for-profit business and a nonprofit organization for a common benefit. Cause marketing is when a company puts a logo on the front of a package, and lets the public know that for every product sold, a percentage of the proceeds go towards supporting the nonprofit. It is a great way for nonprofits to raise funds, create awareness, and get people activated. There are many benefits to companies that support nonprofits, as highlighted by the variety of studies at Engage for Good: www.engageforgood.com/guides/statistics-every-cause-marketer-should-know/

Our partner company, The Mood Factory, did the first nationwide cause marketing campaign for mental health. It is what enabled us to start our work on Hope. We are so grateful for our cause marketing partners, as they have gotten us where we are today.

If you are a brand, join our mission to help teach Hope for free to children around the world, and work with us to create a cause marketing campaign about the "how-to" of Hope. We must teach while we talk. If you are a retailer, consider creating a retail display about Hope in your store and support those that support us.

Your use of Hope branding can help us spread the "how-to" of Hope to people around the world.

Interested in placing Hope branding on your products?
Contact us at activate@theshinehopecompany.com to learn more.

May you plant the seeds, and remember
no matter what darkness comes, there is always a
way to get back to light. Keep Shining Hope.
iFred
shine a light on
depression HOPE
Flap Area (No Copy)
Fold Line

Cause Marketing

International Day of Hope

A celebration of the "how-to" of hope

Join us on July 12th to celebrate the International Day of Hope.

April is National Hope Month. It is therefore fitting that we start the month by celebrating hope, as hope is critically important for mental health. Hope is a known protective factor for anxiety, depression, addiction, self-harm, and suicide.

The International Day of Hope kicks off a 30-Day Shine Hope Challenge dedicated to sharing the science, stories, and strategies of hope around the world.

Learn more about the International Day of Hope by visiting www.internationaldayofhope.org

#HopefulCities   www.hopefulcities.org   Media Sponsored by

International Day of Hope Poster

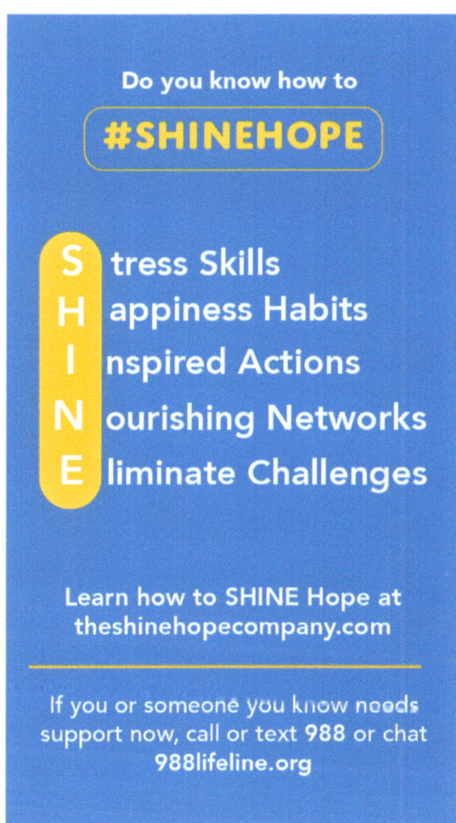

Do you know how to
#SHINEHOPE
S tress Skills
H appiness Habits
I nspired Actions
N ourishing Networks
E liminate Challenges
Learn how to SHINE Hope at
theshinehopecompany.com
If you or someone you know needs
support now, call or text 988 or chat
988lifeline.org

"Once you choose hope,
anything's possible."
- Christopeher Reeve

Moment of Hope Cards

HOPE is Teachable

Resources available at www.hopefulcities.org

Space for Sponsor's Logo

HOPE is Teachable

Resources available at www.hopefulcities.org

got hope?

Find out at www.hopefulcities.org

Do you know how to SHINE HOPE?

Learn how at www.hopefulcities.org

HOPE ≠ WISH

Learn more at www.hopefulcities.org

HOPE is Teachable

Resources available at www.hopefulcities.org

What is your Hope Score?

Find out today at www.hopefulcities.org

Join the Hopeful Cities Movement

Find out how at www.hopefulcities.org

START ACTIVATING HOPE TODAY.

Find out how at www.hopefulcities.org

These are signs for HOPE
measurable, teachable, and free

www.hopefulcities.org

Billboards

HOPE
is teachable
#HopefulCities
www.hopefulcities.org

HOPE
is teachable
#HopefulCities
www.hopefulcities.org

Space for Sponsor's Logo

HOPE
is teachable
#HopefulCities
www.hopefulcities.org

Do you know how to
SHINE
HOPE?
#HopefulCities
www.hopefulcities.org

got hope?
#HopefulCities
www.hopefulcities.org

Join the Hopeful
Cities Movement
#HopefulCities
www.hopefulcities.org

These are signs for
HOPE
measurable, teachable, and free
#HopefulCities
www.hopefulcities.org

These are signs for
HOPE
#HopefulCities
www.hopefulcities.org

Take the
Global Hope
Challenge
#HopefulCities
www.hopefulcities.org

Yard Signs

Join the Hopeful Cities Movement

Higher hope can positively impact work performance, athletic achievement, health, resilience, and more. Learn the "how-to" of hope with the Five-Day Global Hope Challenge.

To find free local resources, and start activating hope in your life and community, visit www.hopefulcities.org. No matter what life brings, there is always a way from hopelessness to hope.

www.hopefulcities.org
#HopefulCities

Hopeful Cities Poster

Whether sharing iconic images of sunflowers for Hope or encouraging budding photographers to encapsulate the power of Hope, photography contests are a great way to #SpreadHope in public spaces. We've run online contests via Viewbug, shown entries at art exhibits, and had photos displayed in local coffee shops.

Art is one of our favorite Stress Skills and Happiness Habits. Creating innovative ways to allow the public to engage in creativity is a great way to activate Hope in your community.

Photography Contest

Mural and Sculpture

International Day of Hope

WHAT IS THE INTERNATIONAL DAY OF HOPE?

- July 12th has been declared the International Day of Hope. The mission of the International Day of Hope is to spread the message that hope is teachable, to share the science, stories, and strategies of hope, and to activate the "how-to" of hope in communities around the world.
- The International Day of Hope kicks off a thirty-day campaign to activate hope in your community through the use of the 30-Day Shine Hope Challenge, yard signs, sunflower gardens, workplace posters, murals, live speaking events, our free online event, teaching Hopeful Minds in the classroom, and so much more!
- As part of the International Day of Hope, Mayors, Governors, and other governmental officials are encouraged to issue a Proclamation making the day official and joining the Hopeful Cities Movement. You can learn more at www.hopefulcities.org.

WHY THE INTERNATIONAL DAY OF HOPE?

- The International Day of Hope is in April because April is National Hope Month. Hope is critically important for mental health, as it is a teachable skill that is a protective factor for anxiety, depression, addiction, self-harm, and suicide.
- iFred has created a Hopeful Cities Playbook, a free tool that cities can use to operationalize "hope" in their own city. The Hopeful Cities Playbook provides instructions about how to become a Hopeful City and implement Day of Hope activities in your community.
- The time has come to join together to create a Global Movement for Hope.

WHY NOW?

Covid-19 and inequality have created unprecedented stress on people around the world. Our ability to effectively manage and adapt to these times determines our success as a society. The Day of Hope will help promote skills for how to proactively manage stress, channel emotions for good, and effectively create change for the better. Join us today.

We would like to thank our Sponsor:

Learn more about the International Day of Hope by visiting www.internationaldayofhope.org

 @ifredorg  @ifredorg  @ifredorg

International Day of Hope Flyer

HOPEFUL CITIES FACT SHEET

"Hope is something we have control over. It's a skill and a motivation, so it is something that we can work towards."
– Myron Belfer, MD, MPA

"Hope is a verb. Hope is action."
– Dr. Edward Barksdale, Jr., MD

Hope Science has been used around the world, and CR Snyder's Hope Scale has been used around the world to measure hope. Snyder's Adult and Child Hope Scales are validated scales and predict life outcomes. These are short surveys found here, and we recommend everyone measure their hope.

theshinehopecompany.com/measure-your-hope/

Higher Hope is associated with:

- Lower levels of violence
- Increased support for addressing climate change
- Lower levels of addiction
- Improved school retention and improved academic performance
- Improved conflict resolution
- Improved workplace productivity, engagement, and retention
- Improved psychiatric and medical outcomes
- Improved goal attainment, which is necessary to progress towards the Sustainable Development Goals (SDGs) set forth by the United Nations.

Do you know how to
SHINE HOPE?

Hopelessness is the key predictor of:

- Violence (4.4 million people are killed each year through violence)
- Addiction (26% rise in substance use across the world since the start of COVID-19)
- Poverty (575 million people are expected to live in poverty by 2030)
- Climate Crisis (the global temperature is expected to reach the 1.5 °C tipping point by 2035)
- Workforce costs (it costs companies $15,000 each year per employee with depression)
- Poor psychiatric and medical health ($2 – 5 trillion per year in global healthcare costs).

HOPE

HOPEFUL CITIES PLAYBOOK
#HopefulCities
#ShineHope

Hopeful Cities teaches how to proactively manage the two components of hopelessness, emotional despair, and motivational helplessness, and how to activate hope through the Five Keys to Shine Hope™: Stress Skills, Happiness Habits, Inspired Actions, Nourishing Networks, and Eliminating Challenges.

Hopelessness is learned, and HOPE is teachable.

www.hopefulcities.org

@theshinehopecompany

Hopeful Cities Fact Sheet

HOPEFUL CITIES FACT SHEET

Our programs are evidence-informed and we are constantly collecting data to demonstrate the programs' effectiveness, while using the data to adapt the programs to different populations.

More findings and sources at www.theshinehopecompany.com/hope-science/general/

On the Hopeful Cities Landing page, you'll find resources for how to hope, including:

- The Hope Scale
- Hopeful Minds Parent's Guide: How to Use Hope Language at Home
- Hopeful Minds Overview Educator Guide and Workbook
- Hopeful Minds Deep Dive Educator Guide and Workbook
- Hopeful Minds for Teens Program
- Shine Hope Infographic
- Digital Shine Hope Brochures
- Social Media Activation
- Workplace Posters
- Where to go for support
- Our aim is to ensure all know the what, why, and how of hope
- Find out more at www.hopefulcities.org

Hope Scales
HOPE
A HOW-TO FOR HOPE IN TRYING TIMES
It's time to choose
HOPE

The programming was developed by Kathryn Goetzko, MBA, and a group of global hope and mental health experts. Kathryn created the first no cost evidence-based program to teach hope around the world through Hopeful Minds, recognized as an innovation by the World Bank. She then went on to create Hopeful Cities, Hopeful Minds, wrote The Biggest Little Book About Hope, is host of The Hope Matrix Podcast, and is CEO of The Shine Hope Company. She is a representative at the United Nations for the World Federation for Mental Health, and is committed to ensuring all around the world have the skills they need to Shine Hope.

www.hopefulcities.org

@theshinehopecompany

Hopeful Cities Fact Sheet

PLANT SUNFLOWER GARDENS TO SHINE HOPE

Gardening is a great time to practice the Shine Hope Framework, as we have a lot of challenges while planting a garden and we can go from hope to hopelessness pretty quickly. Yet that is a normal part of life, so gardening is an easy place to start practicing these skills.

Say you find some tough ground you need to dig into to plant, you may get frustrated and give up. It is a good time to practice a **Stress Skill** like a 90-second pause or deep breathing, to calm down your stress response. Then try again! You may also notice when others get frustrated and teach them how to use this skill to navigate from their downstairs brain back upstairs.

Eating the sunflower seeds (if ok with your doctor) might be a good way for you to practice your **Happiness Habits.** Sunflower seeds are nutritious, high in choline and selenium, great for brain function and memory. You might also get some exercise planting gardens, and spend time in nature, two other Happiness Habits and great ways to release endorphins.

Planting gardens remind us to take **Inspired Actions** by setting specific goals for the garden. If we want a garden, we need to set a SMART goal about how many flowers, when and where we want the garden, and how we are going to grow the flowers. It is best if we write down the plan, chunk it down into actionable steps, think about obstacles and multiple ways we might overcome them, and check in with someone regularly to ensure progress.

We can cultivate our **Nourishing Networks** by planting gardens with others. That way, if we have challenges while planting, we can face them together and be more creative about overcoming them. And if we don't live by the person we want to plant with, we can both decide to plant and check in regularly on the garden. It is also super fun to plan community gardens, or even fields of sunflowers, and all join together in learning and practicing skills to Shine Hope.

And finally, time to get serious about **Eliminating Challenges**. For example, if our sunflowers die and we fail for a season of planting, it is easy for us to think of ourselves as failures. Yet we aren't failures, our process failed. So deconstruct the process. Did we under or over water? Did we plant at the wrong time of year? Was something wrong with the soil? Did we overwater? It is time to investigate, and instead of ruminating about the sunflowers start figuring out what we can do better to try again next year.

Planting sunflowers is a way to spread the message of hope, as if you put up a Gardens of Hope sign with the website, people can then find the curriculum to learn more about the programs for 'how' to hope. Our program is available around the world, and gardens are a great way to share the message that Hope is Teachable.

Find out more at www.hopefulcities.org

@theshinehopecompany    

Gardening Sheet

Shine Hope Products

Links and References

Hopelessness Stats

Violence

World Health Organization. Injuries and Violence. www.who.int/teams/social-determinants-of-health/injuries-and-violence

Addiction

United Nations Office on Drugs and Crime (UNODC). (2022). World Drug Report 2022. reliefweb.int/report/world/unodc-world-drug-report-2022

Mental Health

Chisholm, D., Sweeny, K., Sheehan, P., Rasmussen, B., Smit, F., Cuijpers, P., & Saxena, S. (2016). Scaling-up treatment of depression and anxiety: A global return on investment analysis. The Lancet Psychiatry. pubmed.ncbi.nlm.nih.gov/27083119/

Poor Health

de Faria, D. A., de Souza, É. D., Belo, V. S., & Botti, N. C. L. (2020). Physical pain and Hopelessness in school teenagers. Brazilian Journal of Pain, Oct-Dec 2020. www.scielo.br/j/brjp/a/zL7S45RPfbL8NN7XR8PDN5t/?format=html&lang=en

Everson, S. A., Goldberg, D. E., Kaplan, G. A., Cohen, R. D., Pukkala, E., Tuomilehto, J., & Salonen, J. T. (1996). Hopelessness and risk of mortality and incidence of myocardial infarction and cancer. Psychosomatic Medicine. pubmed.ncbi.nlm.nih.gov/8849626/

Katon, W. J. (2011). Epidemiology and treatment of depression in patients with chronic medical illness. Dialogues in Clinical Neuroscience. www.ncbi.nlm nih.gov/pmc/articles/PMC3181964/

Morgado P and Cerqueira JJ (2018) Editorial: The Impact of Stress on Cognition and Motivation. www.frontiersin.org/articles/10.3389/fnbeh.2018.00326/full

Pedersen, S. S., Denollet, J., Erdman, R. A. M., et al. (2009). Co-occurrence of diabetes and Hopelessness predicts adverse prognosis following percutaneous coronary intervention. Journal of Behavioral Medicine. link.springer.com/article/10.1007/s10865-009-9204-9

Workforce Costs

National Safety Council. (2023). Prioritizing Employee Mental Health. www.nsc.org/workplace/safety-topics/employee-mental-health

Hope Stats

Violence

Stoddard, S. A., McMorris, B. J., & Sieving, R. E. (2011). Do social connections and Hope matter in predicting early adolescent violence? American Journal of Community Psychology. www.ncbi.nlm.nih.gov/pmc/articles/PMC3165137/

Addiction

Brooks, M. J., Marshal, M. P., McCauley, H. L., Douaihy, A., & Miller, E. (2016). The Relationship Between Hope and Adolescent Likelihood to Endorse Substance Use Behaviors in a Sample of Marginalized Youth. www.ncbi.nlm.nih.gov/pmc/articles/PMC8006866/

Poverty

Eggers, A., Graham, C., & Sukhtankar, S. (2004, February 1). Does Happiness Pay? An Exploration Based on Panel Data from Russia. www.brookings.edu/articles/does-happiness-pay-an-exploration-based-on-panel-data-from-russia/

Climate Crisis

Bury, S.M., Wenzel, M. and Woodyatt, L. (2020), Against the odds: Hope as an antecedent of support for climate change action. bpspsychub.onlinelibrary.wiley.com/doi/abs/10.1111/bjso.12343

Displacement

Cohen-Chen, S., Halperin, E., Crisp, R. J., & Gross, J. J. (2014). Hope in the Middle East: Malleability Beliefs, Hope, and the Willingness to Compromise for Peace. journals.sagepub.com/doi/abs/10.1177/1948550613484499?casa_token=1JJg6UGlpdkAAAAA%3A_gijo1Wixy7jcZ8C6mxAzK0go1Gbf-9P8K0zxW5XGbQkuM2M9wmoRNXIuP_MQYWpYbLQg4kZPRQ&journalCode=sppa

Education Crisis

Bashant, J. L. (2016). Instilling Hope In Students. Journal for Leadership and Instruction. files.eric.ed.gov/fulltext/EJ1097567.pdf

Bryce, C. I., Fraser, A. M. J., Fabes, R. A., & Alexander, B. L. (2021). The role of Hope in college retention. www.sciencedirect.com/science/article/abs/pii/S1041608021000704

Hodges, T. (2016, August 31). Student Hope, Engagement as Important as Graduation Rates. news.gallup.com/opinion/gallup/195248/student-Hope-engagement-important-graduation-rates.aspx

Gender inequality

Greenaway, K. H., Cichocka, A., van Veelen, R., Likki, T., & Branscombe, N. R. (2016). Feeling Hopeful Inspires Support for Social Change. http://www.jstor.org/stable/43783897

Health

Long, K. N. G., Kim, E. S., Chen, Y., Wilson, M. F., Worthington Jr, E. L., & VanderWeele, T. J. (2020). The role of Hope in subsequent health and well-being for older adults: An outcome-wide longitudinal approach. www.sciencedirect.com/science/article/pii/S259011332030002X

Zhu AQ, Kivork C, Vu L, Chivukula M, Piechniczek-Buczek J, Qiu WQ, Mwamburi M. (2017)The association between Hope and mortality in homebound elders. www.ncbi.nlm.nih.gov/pmc/articles/PMC5552440/

Zou, T., Liu, J., & Tan, L. (2022). Study on the Effect of Hope Theory Combined with Psychological Intervention on the Improvement of Prognosis. doi.org/10.1155/2022/1153071

Workforce

Weir, K. (2013, October 1). Mission impossible. www.apa.org/monitor/2013/10/mission-impossible

Bareket-Bojmel, L., Chernyak-Hai, L., & Margalit, M. (2023). Out of sight but not out of mind: The role of loneliness and Hope in remote work and in job engagement. doi.org/10.1016/j.paid.2022.111955

Fazal-e-Hasan, S. M., Ahmadi, H., Sekhon, H., Mortimer, G., Sadiq, M., Kharouf, H., & Abid, M. (2023). The role of green innovation and Hope in employee retention. doi.org/10.1002/bse.3126

Hope and Sustainable Development Goals (SDGs)

Moss, S. A. (2018). Hope and goal outcomes: The role of goal-setting behaviors [Master's thesis, Ohio State University]. http://rave.ohiolink.edu/etdc/view?acc_num=osu1513865199503514

Oettingen, G., & Gollwitzer, P. M. (2002). Turning Hope Thoughts into Goal-Directed Behavior.http://www.jstor.org/stable/1448874

Goal 1: End poverty in all its forms everywhere

Patel, V., & Kleinman, A. (2003). Poverty and common mental disorders in developing countries. Bulletin of the World Health Organization, 81(8). www.scielosp.org/article/ssm/content/raw/?resource_ssm_path=/media/assets/bwho/v81n8/v81n8a11.pdf

Ridley, M., Rao, G., Schilbach, F., & Patel, V. (2020). Poverty, depression, and anxiety: Causal evidence and mechanisms. pubmed.ncbi.nlm.nih.gov/33303583/

Eggers, A., Graham, C., & Sukhtankar, S. (2004, February 1). Does Happiness Pay? An Exploration Based on Panel Data from Russia. www.brookings.edu/articles/does-happiness-pay-an-exploration-based-on-panel-data-from-russia/

Goal 2: End hunger, achieve food security and improved nutrition, and promote sustainable agriculture

Gilbert, J. R., & Ashley, C. (2020). Access Granted? An Examination of Financial Capability, Trait Hope, Perceived Access, and Food Insecurity in Distressed Census Tracts. doi.org/10.1177/0743915619889341

Bukchin, S., & Kerret, D. (2018). Food for Hope: The Role of Personal Resources in Farmers' Adoption of Green Technology. doi.org/10.3390/su10051615

Peterson, C., Sussell, A., Li, J., Schumacher, P. K., Yeoman, K., & Stone, D. M. (2020). Suicide Rates by Industry and Occupation — National Violent Death Reporting System, 32 States, 2016. MMWR Morbidity and Mortality Weekly Report, 69, 57–62. http://dx.doi.org/10.15585/mmwr.mm6903a1

Behere, P. B., & Bhise, M. C. (2009). Farmers' suicide: Across culture. www.ncbi.nlm.nih.gov/pmc/articles/PMC2802368/

Huen, J. M. Y., Ip, B. Y. T., Ho, S. M. Y., & Yip, P. S. F. (2015). Hope and Hopelessness: The Role of Hope in Buffering the Impact of Hopelessness on Suicidal Ideation. doi.org/10.1371/journal.pone.0130073

Goal 3: Ensure healthy lives and promote well-being for all at all ages

Harvard T.H. Chan School of Public Health. (2021). Health benefits of Hope. www.hsph.harvard.edu/news/hsph-in-the-news/health-benefits-of-Hope/#:~:text=An%20optimistic%20outlook%20may%20help,other%20chronic%20conditions%2C%20research%20suggests

Berg, C. J., Ritschel, L. A., Swan, D. W., An, L. C., & Ahluwalia, J. S. (2011). The Role of Hope in Engaging in Healthy Behaviors Among College Students. doi.org/10.5993/AJHB.35.4.3

Meraz, R., McGee, J., Ke, W., & Osteen, K. (2023). Resilience mediates the effects of self-care activation and Hope on medication adherence in heart failure patients. pubmed.ncbi.nlm.nih.gov/37076776/

Nsamenang, S., & Hirsch, J. (2015). Positive psychological determinants of treatment adherence among primary care patients. www.cambridge.org/core/journals/primary-health-care-research-and-development/article/positive-psychological-determinants-of-treatment-adherence-among-primary-care-patients/080830815F0298332F62C67DAF1F8CCE

Feldman, D. B., & Sills, J. R. (2013). Hope and cardiovascular health-promoting behaviour: Education alone is not enough. pubmed.ncbi.nlm.nih.gov/23289597/

Long, K. N. G., Kim, E. S., Chen, Y., Wilson, M. F., Worthington Jr, E. L., & VanderWeele, T. J. (2020). The role of Hope in subsequent health and well-being for older adults: An outcome-wide longitudinal approach. doi.org/10.1016/j.gloepi.2020.100018

Senger, A. R. (2023). Hope's relationship with resilience and mental health during the COVID-19 pandemic. www.ncbi.nlm.nih.gov/pmc/articles/PMC9886565/

Worthington Jr., E. L. (2020, June 17). How Hope Can Keep You Happier and Healthier. greatergood.berkeley.edu/article/item/how_Hope_can_keep_you_happier_and_healthier

Javanmardifard, S., Heidari, S., Sanjari, M., Yazdanmehr, M., & Shirazi, F. (2020). The relationship between spiritual well-being and Hope, and adherence to treatment regimen in patients with diabetes. doi.org/10.1007/s40200-020-00586-1

Kurita, N., Wakita, T., Ishibashi, Y., Fujimoto, S., Yazawa, M., Suzuki, T., Koitabashi, K., Yanagi, M., Kawarazaki, H., Green, J., Fukuhara, S., & Shibagaki, Y. (2020). Association between health-related Hope and adherence to prescribed treatment in CKD patients: multicenter cross-sectional study. doi.org/10.1186/s12882-020-02120-0

Long, K. N. G., Kim, E. S., Chen, Y., Wilson, M. F., Worthington Jr, E. L., & VanderWeele, T. J. (2020). The role of Hope in subsequent health and well-being for older adults: An outcome-wide longitudinal approach. doi.org/10.1016/j.gloepi.2020.100018

Zhu, A. Q., Kivork, C., Vu, L., Chivukula, M., Piechniczek-Buczek, J., Qiu, W. Q., and Mwamburi, M. (2017) The association between Hope and mortality in homebound elders. www.ncbi.nlm.nih.gov/pmc/articles/PMC5552440/

Zou, T., Liu, J., & Tan, L. (2022). Study on the Effect of Hope Theory Combined with Psychological Intervention on the Improvement of Prognosis. doi.org/10.1155/2022/1153071

Goal 4: Ensure inclusive and equitable quality education and promote lifelong learning opportunities for all

Bryce CI, Alexander BL, Fraser AM, Fabes RA. (2019) Dimensions of Hope in adolescence: Relations to academic functioning and well-being. doi.org/10.1002/pits.22311

Day, L., Hanson, K., Maltby, J., Proctor, C., & Wood, A. (2010). Hope uniquely predicts objective academic achievement above intelligence, personality, and previous academic achievement. doi.org/10.1016/j.jrp.2010.05.009

Dixson, D. D., & Stevens, D. (2018). A Potential Avenue for Academic Success: Hope Predicts an Achievement-Oriented Psychosocial Profile in African American Adolescents. doi.org/10.1177/0095798418805644

Zakrzewski, V. (2012, November 6). How to Help Students Develop Hope. greatergood.berkeley.edu/article/item/how_to_help_students_develop_Hope

Hjorth, C. F., Bilgrav, L., Frandsen, L. S., et al. (2016). Mental health and school dropout across educational levels and genders: A 4.8-year follow-up study. doi.org/10.1186/s12889-016-3622-8

Bryce, C. I., Alexander, B. L., Fraser, A. M., & Fabes, R. A. (2020). Dimensions of Hope in adolescence: Relations to academic functioning and well-being. doi.org/10.1002/pits.22311

Dixson, D. D., & Stevens, D. (2018). A Potential Avenue for Academic Success: Hope Predicts an Achievement-Oriented Psychosocial Profile in African American Adolescents. doi.org/10.1177/0095798418805644

Idan, Orly & Margalit, Malka. (2013). Hope Theory in Education Systems. www.researchgate.net/profile/Orly-Idan/publication/264933141_HOPE_THEORY_IN_EDUCATION_SYSTEMS/links/53f60b490cf2fceacc6fda58/HOPE-THEORY-IN-EDUCATION-SYSTEMS.pdf

David Halpin (2001) The Nature of Hope and its Significance for Education. doi.org/10.1111/1467-8527.t01-1-00184

Day, L., Hanson, K., Maltby, J., Proctor, C., & Wood, A. (2010). Hope uniquely predicts objective academic achievement above intelligence, personality, and previous academic achievement. doi.org/10.1016/j.jrp.2010.05.009

Goal 5: Achieve gender equality and empower all women and girls.

Centers for Disease Control and Prevention (CDC). (2023). U.S. Teen Girls Experiencing Increased Sadness and Violence. Retrieved from www.cdc.gov/media/releases/2023/p0213-yrbs.html

Amnesty International. (2023). Women's rights are human rights! Retrieved from www.amnesty.org/en/what-we-do/discrimination/womens-rights/#:~:text=But%20across%20the%20globe%20many,to%20education%2C%20and%20inadequate%20healthcare.

Goal 6: Ensure availability and sustainable management of water and sanitation for all

Fritze, J. G., Blashki, G. A., Burke, S., et al. (2008). Hope, despair and transformation: Climate change and the promotion of mental health and wellbeing. doi.org/10.1186/1752-4458-2-13

Rahimipour, M., Shahgholian, N., & Yazdani, M. (2015). Effect of Hope therapy on depression, anxiety, and stress among the patients undergoing hemodialysis. doi.org/10.4103/1735-9066.170007

Goal 6: Ensure availability and sustainable management of water and sanitation for all

Fritze, J. G., Blashki, G. A., Burke, S., et al. (2008). Hope, despair and transformation: Climate change and the promotion of mental health and wellbeing. doi.org/10.1186/1752-4458-2-13

Rahimipour, M., Shahgholian, N., & Yazdani, M. (2015). Effect of Hope therapy on depression, anxiety, and stress among the patients undergoing hemodialysis. doi.org/10.4103/1735-9066.170007

Goal 7: Ensure access to affordable, reliable, sustainable, and modern energy for all

Marlon, J. ., Bloodhart, B. ., Ballew, M. ., Rolfe-Redding, J. ., Roser-Renouf, C. ., Leiserowitz, A. ., & Maibach, E. . (2019). How Hope and Doubt Affect Climate Change Mobilization. http://doi.org/10.3389/fcomm.2019.00020

Chang E. C. (1998). Hope, problem-solving ability, and coping in a college student population: some implications for theory and practice. pubmed.ncbi.nlm.nih.gov/9811132/

Geiger, N., Gasper, K., Swim, J. K., & Fraser, J. (2019). Untangling the components of Hope: Increasing pathways (not agency) explains the success of an intervention that increases educators' climate change discussions. doi.org/10.1016/j.jenvp.2019.101366

Marlon, J. ., Bloodhart, B. ., Ballew, M. ., Rolfe-Redding, J. ., Roser-Renouf, C. ., Leiserowitz, A. ., & Maibach, E. . (2019). How Hope and Doubt Affect Climate Change Mobilization. http://doi.org/10.3389/fcomm.2019.00020

Goal 8: Promote sustained, inclusive, and sustainable economic growth, full and productive employment and decent work for all

Berg, P. (2020, March 30). Remember the Needs of Followers During COVID-19. www.gallup.com/workplace/304607/remember-needs-followers-during-covid.aspx

Lopez, S. J. (2013). Making Hope Happen: Create the Future You Want for Yourself and Others. www.amazon.com/Making-Hope-Happen-Create-Yourself/dp/1451666225/?utm_source=link_newsv9&utm_campaign=item_160361&utm_medium=copy

Hicks, E., & McFarland, C. (2020). Hope as a protective factor for cognitive difficulties during the COVID-19 pandemic. doi.org/10.15761/fwh.1000186

Chodavadia, P., Teo, I., Poremski, D., Fung, D. S. S., & Finkelstein, E. A. (2023). Prevalence and economic burden of depression and anxiety symptoms among Singaporean adults: results from a 2022 web panel. doi.org/10.1186/s12888-023-04581-7

Goal 9: Build resilient infrastructure, promote inclusive and sustainable industrialization, and foster innovation

Weronika, Trzmielewska & Rak, Tomasz & Wrześniowski, Szymon. (2022). Does Hope in Mind Influence People's Problem-Solving Performance? www.researchgate.net/publication/356392893_Does_Hope_in_Mind_Influence_People%27s_Problem-Solving_Performance

Senger A. R. (2023). Hope's relationship with resilience and mental health during the COVID-19 pandemic. doi.org/10.1016/j.copsyc.2023.101559

Goal 10: Reduce income inequality within and among countries

Mitchell, U. A., Gutierrez-Kapheim, M., Nguyen, A. W., & Al-Amin, N. (2020). Hopelessness Among Middle-Aged and Older Blacks: The Negative Impact of Discrimination and Protecting Power of Social and Religious Resources. doi.org/10.1093/geroni/igaa044

Goal 11: Make cities and human settlements inclusive, safe, resilient, and sustainable

Brooks, M. J., Marshal, M. P., McCauley, H. L., Douaihy, A., & Miller, E. (2016). The Relationship Between Hope and Adolescent Likelihood to Endorse Substance Use Behaviors in a Sample of Marginalized Youth. doi.org/10.1080/10826084.2016.1197268

Martin, K., & Stermac, L. (2010). Measuring Hope: Is Hope related to criminal behaviour in offenders? pubmed.ncbi.nlm.nih.gov/19423753/

Mathis, G. M., Ferrari, J. R., Groh, D. R., & Jason, L. A. (2009). Hope and Substance Abuse Recovery: The Impact of Agency and Pathways within an Abstinent Communal-living Setting. doi.org/10.1080/15560350802712389

Dekhtyar, M., Beasley, C. R., Jason, L. A., & Ferrari, J. R. (2012). Hope as a Predictor of Reincarceration Among Mutual-Help Recovery Residents. doi.org/10.1080/10509674.2012.711806

Hill, L. M., Abler, L., Maman, S., Twine, R., Kahn, K., MacPhail, C., & Pettifor, A. (2018). Hope, the Household Environment, and Sexual Risk Behaviors Among Young Women in Rural South Africa. doi.org/10.1007/s10461-017-1945-9

Li, M., Chang, E. C., & Chang, O. D. (2022). Beyond the Role of Interpersonal Violence in Predicting Negative Affective Conditions in Adults: An Examination of Hope Components in Chinese College Students. doi.org/10.1177/0886260520938515

Goal 12: Ensure sustainable consumption and production patterns

Bukchin, S., & Kerret, D. (2020). Once you choose Hope: Early adoption of green technology. doi.org/10.1007/s11356-019-07251-y

Goal 13: Take urgent action to combat climate change and its impacts by regulating emissions and promoting developments in renewable energy

Bury, S.M., Wenzel, M. and Woodyatt, L. (2020), Against the odds: Hope as an antecedent of support for climate change action. doi.org/10.1111/bjso.12343

Ojala M. (2023). Hope and climate-change engagement from a psychological perspective. doi.org/10.1016/j.copsyc.2022.101514

Maria Ojala maria.ojala@edu.uu.se (2012) Hope and climate change: the importance of Hope for environmental engagement among young people. www.tandfonline.com/doi/abs/10.1080/13504622.2011.637157

Sangervo, J., Jylhä, K. M., & Pihkala, P. (2022). Climate anxiety: Conceptual considerations, and connections with climate Hope and action. www.sciencedirect.com/science/article/pii/S0959378022001078?via%3Dihub

Goal 14: Conserve and sustainably use the oceans, seas and marine resources for sustainable development

Costello, C., Cao, L., Gelcich, S., et al. (2020). The future of food from the sea. www.nature.com/articles/s41586-020-2616-y

National Oceanic and Atmospheric Administration (NOAA). (2023). What does the ocean have to do with human health? oceanservice.noaa.gov/facts/ocean-human-health.html#:~:text=Intensive%20use%20of%20our%20ocean,chemical%20pollutants%20are%20other%20signals.

National Oceanic and Atmospheric Administration (NOAA). (2020). Ocean pollution and marine debris. www.noaa.gov/education/resource-collections/ocean-coasts/ocean-pollution#:~:text=All%20marine%20debris%20comes%20from,such%20as%20tsunamis%20and%20hurricanes.

Zakrzewski, V. (2012, November 6). How to Help Students Develop Hope. greatergood.berkeley.edu/article/item/how_to_help_students_develop_Hope

Goal 15: Protect, restore and promote sustainable use of terrestrial ecosystems, sustainably manage forests, combat desertification, and halt and reverse land degradation and halt biodiversity loss

Bukchin, S., & Kerret, D. (2020). Once you choose Hope: Early adoption of green technology. doi.org/10.1007/s11356-019-07251-y

Shira Bukchin Peles & Dorit Kerret (2021) Sustainable technology adoption by smallholder farmers and goal-oriented Hope. www.tandfonline.com/doi/abs/10.1080/17565529.2021.1872477

Wang, S. (2022). The positive effect of green agriculture development on environmental optimization: Measurement and impact mechanism. www.frontiersin.org/articles/10.3389/fenvs.2022.1035867/full#:~:text=Green%20agriculture%20uses%20modern%20science,environmental%20function%20of%20the%20land.

Goal 16: Promote peaceful and inclusive societies for sustainable development, provide access to justice for all and build effective, accountable and inclusive institutions at all levels

Bolland, J. (2003). Hopelessness and risk behavior among adolescents living in high-poverty inner-city neighborhoods. www.researchgate.net/publication/10904471_Hopelessness_and_risk_behaviour_among_adolescents_living_in_high-poverty_inner-city_neighbourhoods

American Psychological Association. (2013, December 1). Gun violence: Prediction, prevention, and policy. www.apa.org/pubs/reports/gun-violence-prevention

Demetropoulos, J. (2017). Hopelessness and Youth Violent Behavior: A Longitudinal Study. scholarsarchive.byu.edu/cgi/viewcontent.cgi?article=7822&context=etd

Cohen-Chen, S., Halperin, E., Crisp, R. J., & Gross, J. J. (2014). Hope in the Middle East: Malleability Beliefs, Hope, and the Willingness to Compromise for Peace. doi.org/10.1177/1948550613484499

Goal 17: Strengthen the means of implementation and revitalize the global partnership for sustainable development.

Merolla, A. J., Bernhold, Q., & Peterson, C. (2021). Pathways to connection: An intensive longitudinal examination of state and dispositional Hope, day quality, and everyday interpersonal interaction. doi.org/10.1177/02654075211001933

Hopelessness is Affecting Your Community

Increased recidivism

Benecchi, L. (2021). Recidivism Imprisons American Progress. Harvard Politics. Retrieved from harvardpolitics.com/recidivism-american-progress/

Dodd, M. A. (2016). Perceptions of factors related to recidivism and recovery (Master's thesis). California State University, Fresno, College of Health and Human Services. Retrieved from scholarworks.calstate.edu/downloads/707958915#:~:text=Many%20factors%20affect%20recidivism%20such,result%20in%20feelings%20of%20Hopelessness.

Increased violence

American Psychological Association. (2013). Gun Violence Prevention: An APA Topical Fact Sheet. Retrieved from www.apa.org/pubs/reports/gun-violence-prevention

Demetropoulos, J. (2017). Hopelessness and Youth Violent Behavior: A Longitudinal Study. Retrieved from scholarsarchive.byu.edu/cgi/viewcontent.cgi?article=7822&context=etd

World Population Review. Gun Deaths by Country. Retrieved from worldpopulationreview.com/country-rankings/gun-deaths-by-country

Increased susceptibility to mental health disorders

Chodavadia, P., Teo, I., Poremski, D., Fung, D. S. S., & Finkelstein, E. A. (2023). Prevalence and economic burden of depression and anxiety symptoms among Singaporean adults: results from a 2022 web panel. BMC psychiatry, 23(1), 104. doi.org/10.1186/s12888-023-04581-7

World Health Organization. (2022). Depression. Retrieved from www.who.int/news-room/fact-sheets/detail/depression

Increased risk of homelessness

Kuo, G. (2019). Yet Another Emerging Global Crisis - Homelessness. Retrieved from mahb.stanford.edu/library-item/yet-another-emerging-global-crisis-homelessness/

Moschion, J., & van Ours, J. C. (2022). Do early episodes of depression and anxiety make homelessness more likely? Journal of Economic Behavior & Organization, 202, 654-674. doi.org/10.1016/j.jebo.2022.08.019

Increased suicide

World Health Organization. (2021, June 17). One in 100 deaths is by suicide. Retrieved from www.who.int/news/item/17-06-2021-one-in-100-deaths-is-by-suicide

Faria, D. A. de ., Souza, É. D. de ., Belo, V. S., & Botti, N. C. L.. (2020). Physical pain and Hopelessness in school teenagers. Brjp, 3(4), 354–358. doi.org/10.5935/2595-0118.20200196

Everson, S. A., Goldberg, D. E., Kaplan, G. A., Cohen, R. D., Pukkala, E., Tuomilehto, J., & Salonen, J. T. (1996). Hopelessness and risk of mortality and incidence of myocardial infarction and cancer. Psychosomatic medicine, 58(2), 113–121. doi.org/10.1097/00006842-199603000-00003

Katon W. J. (2011). Epidemiology and treatment of depression in patients with chronic medical illness. Dialogues in clinical neuroscience, 13(1), 7–23. doi.org/10.31887/DCNS.2011.13.1/wkaton

UNODC. (2022). World Drug Report 2022. Retrieved from reliefweb.int/report/world/unodc-world-drug-report-2022

Hope: Empowering Cities

Moss, S. A. (2018). Hope and goal outcomes: The role of goal-setting behaviors. Retrieved from etd.ohiolink.edu/acprod/odb_etd/etd/r/1501/10?clear=10&p10_accession_num=osu1513865199503514

Oettingen, G., & Gollwitzer, P. M. (2002). Turning Hope Thoughts into Goal-Directed Behavior. Psychological Inquiry, 13(4), 304–307. http://www.jstor.org/stable/1448874

Improved education

Bashant, J. L. (2016). Instilling Hope In Students. Journal for Leadership and Instruction. files.eric.ed.gov/fulltext/EJ1097567.pdf

Curry, L. A., Snyder, C. R., Cook, D. L., Ruby, B. C., & Rehm, M. (1997). Role of Hope in academic and sport achievement. Journal of personality and social psychology, 73(6), 1257–1267. doi.org/10.1037//0022-3514.73.6.1257

Day, L., Hanson, K., Maltby, J., Proctor, C., & Wood, A. (2010). Hope uniquely predicts objective academic achievement above intelligence, personality, and previous academic achievement. doi.org/10.1016/j.jrp.2010.05.009

Curry, L. A., Snyder, C. R., Cook, D. L., Ruby, B. C., & Rehm, M. (1997). Role of Hope in academic and sport achievement. Journal of personality and social psychology, 73(6), 1257–1267. doi.org/10.1037//0022-3514.73.6.1257

Improved work performance

Weir, K. (2013, October 1). Mission impossible. www.apa.org/monitor/2013/10/mission-impossible

Bareket-Bojmel, L., Chernyak-Hai, L., & Margalit, M. (2023). Out of sight but not out of mind: The role of loneliness and Hope in remote work and in job engagement. doi.org/10.1016/j.paid.2022.111955

Fazal-e-Hasan, S. M., Ahmadi, H., Sekhon, H., Mortimer, G., Sadiq, M., Kharouf, H., & Abid, M. (2023). The role of green innovation and Hope in employee retention. doi.org/10.1002/bse.3126

Trezise, A., McLaren, S., Gomez, R., Bice, B., & Hodgetts, J. (2018). Resiliency among older adults: dispositional Hope as a protective factor in the insomnia-depressive symptoms relation. Aging & mental health, 22(8), 1088–1096. doi.org/10.1080/13607863.2017.1334191

Warwick, B.. (2012). Desfecho após alta da unidade de terapia intensiva pediátrica. Jornal De Pediatria, 88(1), 1–3. doi.org/10.2223/JPED.2165

Improved Health

Brooks, M. J., Marshal, M. P., McCauley, H. L., Douaihy, A., & Miller, E. (2016). The Relationship Between Hope and Adolescent Likelihood to Endorse Substance Use Behaviors in a Sample of Marginalized Youth. www.ncbi.nlm.nih.gov/pmc/articles/PMC8006866/

Harvard T.H. Chan School of Public Health. (2021). Health benefits of Hope. www.hsph.harvard.edu/news/hsph-in-the-news/health-benefits-of-Hope/#:~:text=An%20optimistic%20outlook%20may%20help,other%20chronic%20conditions%2C%20research%20suggests

Hill, L. M., Abler, L., Maman, S., Twine, R., Kahn, K., MacPhail, C., & Pettifor, A. (2018). Hope, the Household Environment, and Sexual Risk Behaviors Among Young Women in Rural South Africa. doi.org/10.1007/s10461-017-1945-9

Morgado P and Cerqueira JJ (2018) Editorial: The Impact of Stress on Cognition and Motivation. www frontiersin.org/articles/10.3389/fnbeh.2018.00326/full

Ong, A. D., Standiford, T., & Deshpande, S. (2018). Hope and stress resilience. In M W. Gallagher & S. J. Lopez (Eds.), The Oxford handbook of Hope (pp. 255–284). Oxford University Press. psycnet.apa.org/record/2017-55500-023

Warwick A. (2012). Recovery following injury hinges upon expectation and Hope. Journal of trauma nursing : the official journal of the Society of Trauma Nurses, 19(4), 251–254. doi.org/10.1097/JTN.0b013e31827598f7

Merolla, A. J., Bernhold, Q., & Peterson, C. (2021). Pathways to connection: An intensive longitudinal examination of state and dispositional Hope, day quality, and everyday interpersonal interaction. doi.org/10.1177/02654075211001933

Additional Resources

Hopeful Minds is based on the research that Hope is teachable. The aim is to equip all students, teachers, and parents with the tools they need to define, learn, and grow a Hopeful Mind. The Hopeful Minds curriculums and resources are available for download at
www.Hopefulminds.org/curriculums

The Five-Day Global Hope Challenge is a daily challenge that introduces the Five Keys to Shine Hope that everyone can use to activate Hope within their lives and their community. The challenge is ideal for governments, workplaces, schools, and more. Sign-up today at www.Hopefulcities.org

Friendship Bench's mission is to get people out of kufungisisa - depression & anxiety - by creating safe spaces and a sense of belonging in communities to improve mental wellbeing and enhance quality of life. To learn more and request a bench placed in your area, visit
www.friendshipbenchzimbabwe.org

Karma Box Project is a community initiative allowing people to give non-perishable food, hygiene products, toiletries, and other useful items to those in need. The boxes are filled up with the goods by anyone in the community and someone in need can take items from the box as needed. To learn more, visit
www.karmaboxproject.org

One World Strong Foundation created the ResilienceNet Mobile App, which empowers and provides support to local, regional, and national terrorism prevention practitioners, relevant frontline responders and individual Americans seeking support. To learn more about the One World Strong Foundation and download their app, visit www.oneworldstrong.org/copy-of-how-we-do-it

National Alliance on Mental Illness (NAMI) is America's largest grassroots mental health organization dedicated to building better lives for Americans affected by mental illness. NAMI offers an abundance of resources for those navigating mental illness or for those seeking to learn more.
Find more at www.nami.org/home

Choose Love Movement nurtures safer and more loving communities through next generation essential life skills and character development programs for all stages of life. Choose Love is an evidence-based curriculum that will help students feel safer, learn better, and achieve more! Find out more at
www.chooselovemovement.org

Hope Means Nevada works to eliminate teen suicide and empower Nevada's youth to live Hopeful lives. Find out more at www.Hopemeansnevada.org

One Mind catalyzes visionary change through science, business and media to transform the world's mental health. Find out more at www.onemind.org

Charter for Compassion supports the emerging global movement that brings compassion to life. It is a global network connecting people, cities, grassroots organizers and leaders to each other. It provides educational resources, organizing tools, and avenues for communication. Find out more at
www.charterforcompassion.org

This Hopeful Cities Playbook was made possible by:

Special Acknowledgement from the Author

I'd like to acknowledge the following individuals that have contributed significantly to my understanding of mental health, SDGs, and Hope Science:
Dr. Myron Belfer, MD, MPA, Dr. Edward Barksdale, MD, Dr. Mathew Gallagher, PhD, Dr. Crystal Bryce, PhD, Chris Underhill MBE, Dr. Vikram Patel, PhD, MB, BS, Dr. Shekhar Saxena, MD, FRCPsych, DAB, MRC, Psych, Dr. Chan Hellman, PhD, Dr. Dan Tomasulo, PhD, TEP, MFA, MAPP, Dr. Karen Kirby, PhD, MSc, BSc., SFHEA, C.Psychol., Dr. Jennifer Cheavens, PhD, Dr. Robert Waldinger, MD, Dr. Pamela Collins, MD, MPH, Dr. Harry Minas, FRANZCP, Dr. Gabriel Ivbijaro, MBE, Dr. Nasser Loza, MBBCh, DPM&N, MSC, FRCPhsyc, Dr. Tsuyoshi Akiyama, MD, PhD, Dr. Graham Thornicroft, FRCPsych FMedSci PhD, Kristy Stark, MA, EdM, BCBA, Dr. Gary Belkin, MD, PhD, MPH, Dr. Jim Doty, MD, Douglas Abrams, Mayor Hillary Schieve, Luis Gallardo, Paolo Delvecchio, MSW, Dr. Phil Wang, MD DrPH, Dr. Gard Jameson, PhD, Dr. Steven Hayes, PhD, Dr. Paul Mitchell, PhD, Dr. John Boyd, PsyD, FACHE, Dr. Siti Raudzah Ghazali, PhD, Dr. Maurizio Fava, MD, Kelly O'Donnell, Michele O'Donnell, Carol Graham, Seth Kahan, Dr. Mohammed Bin Hamad Bin J. Al-Thani, Dr. Moitreyee Sinha, Tyler Norris, M.Div., Patricio Marquez, Craig Kramer, Luis Gallegos, Bruce Springsteen, Oprah Winfrey , Michelle and Barack Obama, George W. Bush, Selena Gomez, Lady Gaga, Jeremy Renner, Dr. Michelle Funk, Nabila Makram, Dr. Keith Whitfield, Sandy Marshall, Dr. Antonella Santuccione, Kirsten Straughan, Doug McMillon, Alice Walton, Jim Walton, Lynne Walton, Rob Walton, Sam Walton, Karlee Silver, Miranda Wolpert, Sophie Straughan, Matthew Jackman, Chantelle Booysen, John Boyd, Psy.D., MHA, Barbara Van Dahlen, Ph.D., Nigel Frith, Dr. Guy Winch, Ph.D., Dr. Delaney Ruston, M.D., Dr. Elizabeth Lombardo, Kimberley Blaine, MA, MFT, Dave Opalewski, Marie Dunne, Nancy Tamosaitis, Dr. Daniel Amen, Scarlett Lewis, Anna Montances, Mic Fariscal, Naneth Jumawid, Veronica O'Brien, the late Anna Unkovich, The Carter Family, Katherine Ponte, BA, JD, MBA, CPRP, Iain Francis Tulley, Dr. Julian Eaton, Walter Harris, Dévora Kestel, Deb Houry, MD, MPH, Sara Bareilles, The Killers, Bradley Cooper, Thomas Lennon, Sylvester Stallone, Taylor Swift, Beyonce, Eminem, John Krasinski, Dwayne "The Rock" Johnson, Keanu Reeves, Serena Williams, BTS, Ed Sheeran, Malala Yousafzai, Magic Johnson, Jane Goodall, Mr. Rogers, Arianna Huffington, John McCain, Zak Williams, Melinda Gates, Shaka Senghor, Ritu Riyat, MPH, David Makram Bishai, Gino Yu, Rohan Dixit, Ahmed Hassoon, MD, Dr. Rachel S. Herz, PhD, Dr. Erwin Benedict Valencia, DPT, MJ Gottlieb, Erin Michaela Brandt, MS, Kelly Davis, Martin Rafferty, Dr. Diane Dreher, PhD, Lian Zeitz, Dr. David B. Feldman, Ph.D, Abigail Johnson, MPH, Sarah Fader, Garen Staglin, Brandon Staglin, Lisa Gordon, Jeff Gordon, Fred Goetzke, Arnold Goetzke, The Carter Family, Damian Kitson, Dr. Sheikh Mohammed, Jamie Kelly, Jamie Davidson, Randy Anderson, RCPF, LADC, Charlene Sunkel, Jagannath Lamichhane, Chloe Hadjimatheou, Jim Foorman, Skip Simpson, Grant Denton, and Shannon Ellis.

I'd like to also acknowledge the brave individuals who made the case for 'why' we need an International Day of Hope in 2023, *including Zoya Awan, Dr. Myron Belfer, MD, MPA, Dr. Crystal I. Bryce, PhD, Kristy L. Stark, M.A., Ed.M., BCBA, Scott Mandell, MBA, Marisa Hamamoto, Chef Grace Ramirez, Sadhvi Bhagawati Saraswati, PhD, Steve Shell, MBA, Emily Ladau, Alex Garfin, Zanade Mann, MaCherie Dunbar, Dr. Edward Barksdale, MD, Maryn Ryan Soref, Tom Dean, and Kelly Dennis, Esq.*

In special gratitude to my mom, two brothers, nieces, and nephew, who made this work possible. And all the students and youth around the world who worked with us to help understand, teach, and activate Hope globally, and the so many that have helped along the way.

This Playbook is in honor of the late Dr. Shane Lopez, a pioneer in Hope science, and my late dad Jon Goetzke, who through his brilliant business teachings, along with Sam Walton, Paul Carter, made this work possible.

About The Shine Hope Company

Our mission is to improve lives globally by teaching scientifically informed and evidence-based methods to measure and cultivate Hope. Learn how to activate Hope in your life and community at www.theshinehopecompany.com.

About iFred

iFred, a 501(c)3 organization established in 2004, worked to shine a positive light on mental health and eliminate stigma through prevention, research and education and created a shift in society's negative perception through positive imagery, rebranding, celebrity engagement, cause marketing campaigns, and establishing the sunflower and color yellow as the international symbols for Hope. iFred worked with The Mood Factory to do the first nationwide cause marketing campaign for mental health in the US, and created the first ever program to teach Hope, based on research it is a teachable skill.For more information, visit www.ifred.org.

www.ingramcontent.com/pod-product-compliance
Lightning Source LLC
Chambersburg PA
CBHW061148030426
42335CB00003B/153

* 9 7 8 1 9 5 4 5 3 1 1 5 4 *